T0146069

NEW FORCES AT WORK
IN MINING

Industry Views of Critical Technologies

D. J. Peterson

Tom LaTourrette

James T. Bartis

Supported by the
Office of Science and Technology

RAND

SCIENCE AND TECHNOLOGY POLICY INSTITUTE

The research described in this report was conducted by RAND's Science and Technology Policy Institute, under Contract ENG-9812731.

Library of Congress Cataloging-in-Publication Data

Peterson, D. J.
    New forces at work in mining : industry views of critical technologies / D.J. Peterson,
Tom LaTourrette, James T. Bartis.
      p. cm.
    "MR-1324-OSTP."
    Includes bibliographical references.
    ISBN 0-8330-2967-3
    1. Mining engineering—Technological innovations—United States. I. LaTourrette,
Tom, 1963– II. Bartis, James T., 1945– III. Title.

TN23 .P48 2001
622—dc21

                                                 00-054742

RAND is a nonprofit institution that helps improve policy and decisionmaking through research and analysis. RAND® is a registered trademark. RAND's publications do not necessarily reflect the opinions or policies of its research sponsors.

Cover design by Maritta Tapanainen (based on a template by Peter Soriano)

Published 2001 by RAND
1700 Main Street, P.O. Box 2138, Santa Monica, CA 90407-2138
1200 South Hayes Street, Arlington, VA 22202-5050
RAND URL: http://www.rand.org/
To order RAND documents or to obtain additional information, contact
Distribution Services: Telephone: (310) 451-7002; Fax: (310) 451-6915; Internet:
order@rand.org

This report presents an overview of technologies critical to the economic health of the mining and quarrying industries in the United States. The findings were derived from confidential discussions with leading representatives of 58 mining firms, equipment manufacturers, research institutions, and other organizations who were selected for their prominent positions in the industry and their ability to think broadly and creatively about current technology trends.

The study was commissioned by the National Institute for Occupational Safety and Health (NIOSH) of the Centers for Disease Control and Prevention. Additional funding was provided by the U.S. Department of Energy (DOE) Office of Industrial Technologies, with organizational support from the White House Office of Science and Technology Policy (OSTP).

This report is intended to

- Aid federal officials in understanding emerging critical-technology trends in the mining industry.

- Help decisionmakers develop priorities for the national mining health and safety research agenda.

- Assist decisionmakers in refining federal funding priorities for mining technology research and development.

The report will also help mining industry executives, researchers, and stakeholders obtain a broad view and understanding of technology trends pertinent to their operations. While mining and quarrying operations share many technologies and processes, the industry is highly fragmented and competitive, and communication among firms is limited. Thus, industry decisionmakers can benefit from their colleagues' experiences and forward-looking perspectives presented herein.

This research builds on two previous RAND technology studies:

- Steven W. Popper, et al., *New Forces at Work: Industry Views Critical Technologies,* MR-1008-OSTP, RAND, 1998.

- Susan Resetar, *Technology Forces at Work: Profiles of Environmental Research and Development at DuPont, Intel, Monsanto, and Xerox,* MR-1068-OSTP, RAND, 1999.

THE RAND SCIENCE AND TECHNOLOGY POLICY INSTITUTE

Originally created by Congress in 1991 as the Critical Technologies Institute and renamed in 1998, the Science and Technology Policy Institute is a federally funded research and development center sponsored by the National Science Foundation and managed by RAND. The Institute's mission is to help improve public policy by conducting objective, independent research and analysis on policy issues that involve science and technology. To this end, the Institute

- Supports the Office of Science and Technology Policy and other Executive Branch agencies, offices, and councils.

- Helps science and technology decisionmakers understand the likely consequences of their decisions and choose among alternative policies.

- Helps improve understanding in both the public and private sectors of the ways in which science and technology can better serve national objectives.

Science and Technology Policy Institute research focuses on problems of science and technology policy that involve multiple agencies. In carrying out its mission, the Institute consults broadly with representatives from private industry, institutions of higher education, and other nonprofit institutions.

Inquiries regarding the Science and Technology Policy Institute may be directed to

Bruce Don
Director, Science and Technology Policy Institute
RAND
1200 South Hayes Street
Arlington, VA 22202-5050
Phone: (703) 413-1100 x5351
Web: http://www.rand.org/centers/stpi
Email: stpi@rand.org

CONTENTS

FIGURES

TABLES

Over the past century, technology advances have had major impacts on mining practices and the nature of the mine and quarry sites in the United States. The evolution of current technologies, as well as the introduction of innovations, will continue and perhaps accelerate in the new century. Several industry objectives will drive future technology change, including

- Lowering production costs.

- Enhancing the productivity of workers and equipment.

- Opening up new reserves and extending the life of existing ore bodies.

- Continuing to meet regulatory and stakeholder requirements in areas such as health and safety, environmental impacts, and land use.

Between March and July 2000, the RAND Science and Technology Policy Institute conducted a series of in-depth, confidential discussions with key members of the mining community to elicit a wide range of views on technology trends in all sectors of the U.S. mining and quarrying industry. The discussions included 58 organizations engaged in coal, metals, aggregates, and industrial minerals production, as well as technology providers and research institutions.

This report brings to light those technologies viewed by industry leaders as critical to the success of the industry currently, and critical technologies likely to be implemented between now and 2020.

THE MINING INDUSTRY

During the discussions, several important trends impacting the mining industry in the United States were highlighted. These include historically low commodity prices, consolidation and globalization, and regulatory and political constraints on new development. According to industry managers, these trends have resulted in a mining community that is risk-averse and struggling against thin profit margins; in addition, the community has severely cut back on technology research and development (R&D). This has led to a situation in which technology investments tend to focus on short-term and incremental refinements rather than long-term or breakthrough innovations.

Given the higher unit prices of the goods they produce, metals producers tend to be investing more heavily in complex and advanced technologies such as dispatch systems, high-precision equipment positioning, and super-size machinery. In the future, metals producers are likely to spearhead the implementation of remote-controlled and automated equipment. Coal producers, on the other hand, presented themselves as less concerned about mine-productivity-enhancing technologies, primarily because of chronic oversupplies on coal markets and regulatory pressures on coal utilization.

The stone and aggregates industry is in a much different position: Quarries have been operating at record levels of output and enjoying strong revenue streams since the early 1990s as a result of the prolonged U.S. economic expansion and generous public-sector infrastructure spending. Given the aggregates industry's financial prospects, as well as its rapid consolidation and less-advanced technology baseline (in comparison with the rest of the mining industry), we conclude that quarries in the United States may see the fastest pace of technological change in the industry in the coming decades. As rising buyers of new machinery and equipment, stone and aggregates producers also stand to become important drivers of mining technology innovation.

THE CRITICAL TECHNOLOGIES

When asked to identify the technologies critical to resolving the major mine-productivity bottlenecks, industry representatives with whom we spoke identified a fairly consistent set of priority areas:

1. Information and communications technologies for process optimization.
2. Remote control and automation.
3. Operations and maintenance.
4. Unit-operations capabilities.

These four priority technology areas are summarized below.

Information and Communications Technologies

Our discussions of critical technologies indicated that the information revolution is coming to the mining industry and will have a significant impact on mine operations in the coming years. Information technologies (IT) were cited frequently as one of the most important advances shaping mining and quarry-ing practices, since they enable both management and staff to monitor, evalu-ate, and adjust operations in real time to maximize productivity and minimize cost.

Mining equipment is increasingly being outfitted with sensors and information-processing capabilities to control and manage operations. As such advanced equipment becomes more widespread, more operations at the mine site can be tied together by communications and data networks, enabling minewide pro-cess integration and control capabilities. Ultimately, these networks can supply data to a central control or to off-site providers who can then "layer-in" a range of services and support functions, such as mine planning and equipment-maintenance solutions.

According to mining executives, the IT revolution is beginning to have a signifi-cant impact at mine sites. They cited several examples:

- Planning and visualization technologies permit mines to accurately simu-late different choices of initial mine design, operations, equipment types, expansion options, closure outcomes, and the ways these factors influence each other.

- Dispatch systems using the Global Positioning System (GPS) monitor equipment positions, direct materials flows, and optimize capacity utiliza-tion in real time.

- GPS-based surveying can now be integrated with high-precision drilling and earthmoving, so mine maps and plans can be updated in real time as materials are moved.

- The availability of sitewide information-sharing has provided the capability to begin integrating previously separate operations, such as surveying, min-ing, processing, and reclamation.

While mine operations are generating more data, many discussants noted that the information is rarely well utilized at the present. Accordingly, another criti-cal technology is effective knowledge management: tools and capabilities for distilling complex mine information into an actionable format that can be comprehended and acted upon in real time.

Remote Control and Automation

Remote control and automation have been high on the mining R&D agenda for a long time; however, there is a lack of consensus within the mining community regarding the desire for, expectations of, and impacts from the deployment of remote-controlled and autonomous equipment. There was no consensus

Table S.1

Anticipated Availability of Autonomous Mining Equipment

Equipment	Anticipated Date
Autopositioning shovel bucket	2001
Autonomous surface-haul truck	2002–2005
Autonomous LHD vehicle	2005
Autonomous surface drill	2003
Autonomous shovel	2005

Source: RAND discussion participants.

among the RAND discussion participants on the feasibility or benefits of remote and autonomous equipment operation, and the discussants also differed in their views on where these technologies would be commercialized most rapidly.

Remote and autonomous control technologies are still in the formative stages and are currently available in only a few specific tasks, such as remote guidance of load-haul-dump (LHD) vehicles and continuous miners, operator-assisted drill positioning, excavator scooping, and vehicle tramming. Despite this climate of uncertainty, study participants expressed the opinion that the requisite technologies are nearing commercial availability for several pieces of machinery (see Table S.1).

Operations and Maintenance

Given the large expenditures on capital equipment characteristic of mining, technology developers and users alike place a very high priority on improving equipment performance and availability through better operations and maintenance practices and technologies. This coincides with assessments of critical technologies across U.S. industry.

Improving equipment operations and maintenance (O&M) has gained greater importance as margins have been squeezed by competition and weak commodity prices, as mining equipment and geological conditions have become more complex, and as mine processes are becoming more tightly integrated.

Several priority O&M concepts currently are being developed and applied in mining, with the central goals of better understanding maintenance requirements, optimizing the use of maintenance resources, and boosting equipment availability:

* Mining equipment is being outfitted with an increasing variety of on-board sensors for monitoring critical systems. Together with enhanced off-board diagnostics, such as vibration analysis, vital-signs monitoring helps operators predict equipment failures and schedule maintenance actions.

- Mining operations are investing in better maintenance areas, greater contamination control, more thorough and effective record-keeping, and more complete and careful work.

- More robust engineering and materials, such as improved lubricants and maintenance-free systems, together with improved service access and "hot-swappable" components, are extending operational capabilities and reducing downtimes.

- Maintenance outsourcing is becoming increasingly prevalent in mining, mirroring a widespread trend in business. Arrangements include shipping equipment to maintenance and repair specialists off-site, transmitting equipment diagnostic data directly to service contractors and parts suppliers, and having contract maintenance personnel and equipment on-site.

Understanding and optimizing equipment operations and availability were cited by several industry leaders as critical prerequisites for the successful implementation of remote and autonomous operations.

Unit-Operations Capabilities

Central to the mining process, unit-operations (unit-ops) machinery and equipment are by definition critical technologies. However, industry representatives noted that despite their importance to the mining process, existing unit-ops equipment technologies are unlikely to change substantially in the next two decades, and with few exceptions, those technologies in use today at mine sites in the United States will still be in use in 2020.

Technology changes that are being implemented are incremental and typically focus on increasing batch size, reducing cycle intervals, and boosting equipment availability. The expected commercial introductions of several unit-ops technologies are summarized in Table S.2.

Table S.2

Anticipated Availability of Selected Unit-Ops Innovations

Technology	Anticipated Date
Solid-state programmable blast detonators	2000
Six-unit miner-bolters	2000
Mechanical cutter for hard-rock applications	2003
Fuel-cell-powered underground equipment	2010
1000-ton-capacity haul truck	2020
150-cubic-yard-capacity shovel	2020

Source: RAND discussion participants.

Many discussants predicted and welcomed the continuation of the trend toward higher-capacity haul trucks, shovels, loaders, and excavators. Some executives, however, questioned whether the size of haul trucks and excavators has reached a feasibility threshold where the economies of scale have peaked.

HUMAN FACTORS

The term *technology* includes not only physical hardware, but also operational procedures, organizational structures, and management practices. The inclusive nature of this definition is important: According to the industry leaders with whom we spoke, some of the most important innovations concern the organization and management of mining and quarrying facilities. To this point, a critical component in the technology equation raised repeatedly in the discussions was human factors.

One major issue on which human factors entered into the technology discussion is health and safety. Interestingly, all things considered, complying with health and safety regulations *was not* cited as a driver of technological change. Rather, industry leaders typically pointed to what they saw as innovative personal safety equipment or programs their firms had undertaken on a voluntary basis that went beyond regulatory requirements. Many innovations cited in the discussions specifically address the perceived need to create a more enjoyable, interesting, and productive work environment in order to attract and retain the best workers in a very competitive hiring environment.

Indeed, the discussion participants suggested that as technology progresses, workers are becoming *more critical* to the success of mining and quarrying operations, not less:

- As mining equipment increases in scale and staffing levels decline, individual operators play a greater role in determining mine output.

- As mining equipment becomes more advanced through IT and communications innovations, line workers are gaining unprecedented access to information and control over the equipment they are operating.

- Achieving the productivity gains sought by both management and investors requires that miners develop new, multidisciplinary skills to fully utilize emerging technologies, and that their roles be upgraded from following rules to solving problems.

At the same time, many participants explicitly downplayed the importance of hardware innovations in determining performance outcomes. Rather, they emphasized the importance of engaging and motivating their workforces. Even when a new technology may be seen as potentially beneficial, building workers' acceptance of the technology and their commitment to using it to its greatest extent is an essential prerequisite to successful implementation.

IMPLICATIONS

Several implications can be drawn from the discussions on technology trends and public- and private-sector involvement in technology research, development, and diffusion in the future.

Discussants from all corners of the mining community agreed that both the amount of activity and the level of cooperation on mining technology R&D efforts in the United States had decreased substantially over the past few decades. Yet discussion participants repeatedly raised the concern that more R&D overall *and* more R&D collaboration among technology suppliers and between technology suppliers and operating companies were required. The need for collaboration is clear: The demand for mining equipment and services is limited, advanced technologies are increasingly complex, and R&D needs are often too great for a single company to support. In short, R&D roles as well as funding levels need to be reappraised in the industry to meet the technology objectives viewed as important by the industry itself. Funding for R&D partnerships provided by the Department of Energy (DOE) Office of Industrial Technologies and supported by the National Mining Association was mentioned as a positive step in this direction—suggesting that the federal government has an important role to play in convening parties and catalyzing ventures.

When asked to highlight critical technologies, operating company participants often focused first on downstream technologies (e.g., for refining) rather than upstream activities such as blasting. This tendency reflects several factors, including the fact that productivity gains tend to increase as value is added to a product while it moves downstream. However, as processing plants become more highly tuned and as the just-in-time delivery principle becomes more important, the quality and stock of feed materials become more important determinants of plant performance. This suggests that R&D and innovations targeted at upstream mining processes are likely to have higher payoffs in the future and thus merit closer attention.

Mining in the United States has a tradition of self-reliance: Mines typically are in remote locations, and the ore bodies they work often have unique characteristics. The discussion participants suggested that existing information transfer and technology search mechanisms in the industry also do not appear to be fully exploiting opportunities for technology crossover within companies, across the mining industry, or, importantly, with nonmining sectors. Previous RAND research has indicated that maintenance is a critical-technology concern for many industries. Diagnostics such as vibration, lubricants, and ultrasonic analysis are now common in industries with relatively small capital investments in their fleets. Over-the-road trucking, construction, and manufacturing—like mining—have high downtime costs, and the mining companies could gain tremendous benefit from learning about new maintenance concepts and best practices in these sectors. The mining industry can also draw on the expertise of the U.S. Department of Defense (DoD), for example, in managing complex maintenance and repair operations and in the advanced use of modular designs, subsystem replacement, and centralized repair depots. Other technology crossover areas of opportunity identified include Total Quality Management (TQM), "Six Sigma" enterprise process redesign, and pollution prevention.

We encountered widely differing views on the prospects for many new technologies, for example, super-size haul trucks and automation. These differences of opinion were rooted in varying assessments of the expected performance, costs, and benefits of new technologies. At the same time, it was repeatedly noted that operating companies—especially smaller ones—have limited knowledge of costs across their operations. This may be skewing priorities in favor of conventional technologies and practices and slowing the development and diffusion of important new technologies. Measuring economic productivity benefits of new technologies—especially IT—is a notoriously difficult task for any industry. Moreover, mining and quarrying operations today also compete on a range of other parameters that are difficult to measure: the ability to attract skilled and motivated workers; health and safety, environmental, and aesthetic impacts; customer satisfaction; and community acceptance. Objective, third-party measurement and assessment of emerging technologies on productivity and other important measures could better support the positive impacts of critical-technology acquisition, as well as R&D investment decisions.

ACKNOWLEDGMENTS

The authors wish to thank all of the individuals who participated in the RAND technology discussions for their time, candor, and good will. We also thank Tim Arnold, Barry Turley, David Zatezalo, and Paul Zerella for graciously arranging facility tours.

We would like to extend special thanks to RAND consultant J. Allen Wampler for his guidance and assistance during the discussions phase of this project, as well as his valuable comments as an external peer reviewer of the draft report. Pamela L. Tomski, also a RAND consultant, played an important role in arranging the industry discussions and gathering background information. Clifford Grammisch provided useful guidance on presenting our arguments. Susan Resetar and Sally Sleeper served as the external peer reviewers for this report and provided background on trends in other sectors and other valuable insights that strengthened our presentation.

This research was greatly facilitated by the interest and support of the National Mining Association and its President and Chief Executive Officer, General Richard Lawson, Ret.; its Senior Vice President for Policy, Constance D. Holmes; its Vice President for Safety and Health, Bruce Watzman; and staff members Joseph Lema and Lisa Corathers. We also appreciated the interest of William C. Ford, Senior Vice President of the National Stone Association. Key organizational backing was provided by Henry Kelly of the White House Office of Science and Technology Policy.

George Bockosh of the National Institute for Occupational Safety and Health conceived and sponsored this research, and Toni Maréchaux of the U.S. Department of Energy Office of Industrial Technologies contributed important supplemental funding. The authors wish to thank both of these individuals for their enthusiastic and collegial support.

While the content of the report reflects the observations and opinions of the people interviewed, the authors accept the responsibility for the ways those views are expressed in these pages.

ACRONYMS AND ABBREVIATIONS

DoD	U.S. Department of Defense
DOE	U.S. Department of Energy
GIS	Geographical Information System
IT	Information technologies
GPS	Global Positioning System
LHD	Load-haul-dump vehicle
NIOSH	National Institute for Occupational Safety and Health
NO_x	Nitrogen oxide
OSTP	Office of Science and Technology Policy
O&M	Operations and maintenance
R&D	Research and development
Unit-ops	Unit operations

Cable bolt	A rock bolt consisting of prestressed, multistrand steel cable. Used for supplemental and primary roof support in underground mines. See *rock bolt*.
Continuous miner	A mobile apparatus for cutting and extraction of coal, consisting of a cylindrical cutting head and a mechanism for transporting the coal to the rear of the machine. Used in underground coal mines for primary coal production and driving entries for longwall panels.
Cyanide	A chemical compound (KCN or NaCN) that complexes strongly with gold. Used as a leaching agent to extract gold from mined ores.
Dragline	The largest class of earthmoving equipment used in mining, consisting of a large bucket suspended by cables from a boom affixed to a rotating house. Used for removing overburden in surface mines, primarily coal.
Drift	The access tunnel in underground metal mines.
Entry	The access tunnel in underground coal mines.
Geographic Information System (GIS)	Mapping software designed for storage and manipulation of spatial data (i.e., an object's location and features) and typically viewed in a graphic format as data layers or overlays.
Global Positioning System (GPS)	An array of 24 satellites that transmit data received by ground units to enable users to determine their position in three dimensions on the earth's surface.
Ground control	The process of stabilizing the roof, walls, and floor of underground mine tunnels and chambers by means of mechanical supports such as hydraulic shields, shotcrete, or bolts.
Jumbo	A drilling apparatus used for drift development in underground metal mines.

Leaky feeder   An underground mine communication system consisting of a coaxial antenna cable distributed throughout the mine and connected to the surface.

Longwall   An underground coal-mining technology in which the coal is extracted by a rotating shearing drum running on a rail parallel to the coal face. A chain-driven conveyor belt below the shearer removes the broken coal. As the assembly advances, the roof above the mined portion behind the assembly is allowed to collapse.

Miner-bolter   A mobile apparatus consisting of an integrated continuous miner and roof- and rock-bolting machines. Used in underground coal mines for primary coal production and driving entries for longwall panels. See *continuous miner*.

Muck   Blasted material; mucking is the process of loading and hauling muck.

Operator-assist   Technology in which particular subtasks in equipment operation, such as shovel scooping or continuous miner guidance, are controlled automatically by a computer.

Quarry   A surface mine producing any kind of sand, gravel, stone, or aggregates.

Rock bolt   A steel rod, from four to 12 feet long and up to one inch in diameter, inserted into open rock faces for roof and wall support in underground mines. Bolts are secured by mechanical anchors, epoxy resin, or grout.

Shotcrete   Concrete applied as a spray-coating; a form of ground control in underground mines.

Stope   A chamber blasted within the ore body in an underground metal mine from which ore is extracted.

Tomography   A technique for imaging solids by analyzing the way seismic waves propagate through different regions, such as ore bodies.

Unit operations   The series of steps involved in ore extraction, traditionally
 (unit-ops)   consisting of drilling, blasting, loading, hauling, and dumping.

Chapter One

INTRODUCTION

Over the past century, technological advances such as jackleg drills, longwall shearers, diesel power, and hydrometallurgy have had major impacts on mining and quarrying practices and the nature of the mine site in the United States.

The evolution of current technologies, as well as the introduction of new, breakthrough technologies, will continue and perhaps accelerate in the new century. Several industry objectives will drive future technology change, including

• Lowering production costs.

• Enhancing the productivity of workers and equipment.

• Opening up new reserves and extending the life of existing ore bodies.

• Meeting regulatory and stakeholder requirements in areas such as health and safety, environmental and aesthetic impacts, and land use.

Many technologies have been proposed for use in U.S. mines, including mechanical cutting of hard rock, remote-controlled and automated machinery, wireless communications and data networks, and in-situ processing. Which technologies will prove to be critical to the success of America's mining industry, and which technologies will have a less important role?

Mining and quarrying are capital-intensive activities, and many factors affect the pace of diffusion of new technologies: research and development (R&D) budgets, commodities markets and profit margins, regulatory and community requirements, the ability of firms to acquire and assimilate information, available technology options, the variability in cost structures among firms, and industry attitudes. In recent years, the organization of the mining industry has changed substantially as a result of enterprise restructuring, consolidation, and globalization. How will these factors drive or inhibit technology change in the coming years?

Drawing on the views of industry leaders, this study addresses these important issues shaping the technology profile of mining in the United States.

ABOUT THE STUDY

Task and Purpose

In 1999, the RAND Science and Technology Policy Institute was requested to conduct a series of in-depth, confidential discussions with key members of the mining community to elicit a wide range of views on technology trends in the U.S. mining industry. The goal of this research was to identify those technologies viewed by industry leaders as critical to the success of their operations currently, as well as technologies likely to be implemented between now and 2020.

This research was commissioned by the National Institute for Occupational Safety and Health (NIOSH) of the Centers for Disease Control and Prevention. In April 1996, NIOSH, along with its various partners, established the National Occupational Research Agenda, a framework to guide occupational safety and health research into the next decade—not only for NIOSH, but also for the entire occupational safety and health community.[1] This process resulted in a consensus about the top 21 research priorities, one of which was "emerging technologies" in the workplace.

In 1997, mining and quarrying firms in the United States employed about 240,000 people, three-quarters of whom were directly involved in production. Advances in technologies at mine and quarry sites provide important opportunities to minimize the drudgery of work and eliminate old hazards, but they also may create new, unforeseen risks to workers. Traditionally, hazards in the mining industry have been addressed retrospectively—that is, after they have occurred—but NIOSH policy also is to anticipate the potential health and safety consequences, both good and bad, of emerging technologies so that interventions can be engineered before accidents occur. This study is intended to define what are likely to be critical emerging technologies, in the opinion of leaders in the mining industry, so that NIOSH can then define the appropriate occupational research agenda to ensure the best possible safety and health outcomes for miners in the future.

Additional support for this study was provided by the Mining Industry of the Future program—a partnership of the U.S. Department of Energy (DOE) Office

---

[1]Information about the National Occupational Research Agenda can be found at http://www.cdc.gov/niosh/norhmpg.html.

of Industrial Technologies and the National Mining Association.[2] An important activity of this partnership has been the industry's development of visions of its future in the United States and technology roadmaps to identify critical pathways for the R&D needed for mining to reach its productivity and other strategic goals. This report is intended to aid both industry and government in making decisions to support R&D that is critical to attaining the industry's vision.[3]

Several other agencies with significant interests in mining and mining-related activities—e.g., the U.S. Mine Safety and Health Administration, the U.S. Minerals Management Service, the U.S. Forest Service, the U.S. Office of Surface Mining, the U.S. Bureau of Reclamation, the U.S. Bureau of Land Management, and state mining departments—can draw insights from these industry discussions. Finally, the U.S. Department of Defense (DoD), with its interests in heavy construction and working in underground environments, may also benefit from the insights presented herein.

Just as important, this report also should help mining-industry executives, researchers, and stakeholders to obtain a broad view and understanding of technology trends pertinent to their operations. While mining and quarrying operations share many technologies and processes, the industry is highly competitive and fragmented, and, according to several industry executives with whom we spoke, information sharing among firms is very limited. Thus, industry decisionmakers can benefit from their colleagues' past experiences and forward-looking perspectives presented herein.

How the Study Was Conducted

This mining study was modeled after the RAND Science and Technology Policy Institute's 1998 National Critical Technologies Assessment and other critical technologies studies.[4]

---

[2]For background on the Mining Industry of the Future program, see http://www.oit.doe.gov/mining/.

[3]We note that RAND was not tasked with identifying priorities for mining-technology research. While the RAND project was under way, a parallel analysis of mining and mineral processing-technology needs was commissioned by the Mining Industry of the Future partnership and was being conducted by the Committee on Technologies for the Mining Industries under the auspices of the National Research Council. It was anticipated that this committee would make recommendations for R&D funding priorities.

[4]Steven W. Popper et al., *New Forces at Work: Industry Views Critical Technologies*, MR-1008-OSTP, RAND, 1998; Susan Resetar, *Technology Forces at Work: Profiles of Environmental Research and Development at DuPont, Intel, Monsanto, and Xerox*, MR-1068-OSTP, RAND, 1999.

RAND researchers met with and led structured discussions with representatives from 58 organizations across a broad spectrum of the mining industry (Figure 1.1). The participating firms included 28 machinery, equipment, and service providers, ranging from heavy equipment manufacturers to suppliers of adhesives and engineering services; eight coal producers; 11 metals (including gold, silver, copper, nickel, iron) producers; and four aggregates and industrial minerals producers. Finally, we spoke with representatives from seven research and government institutions that have important roles in the technology development and demonstration process.

The majority of the participants were drawn from the executive ranks of the mining industry; they include chief executive officers, presidents, chief operating officers, and vice presidents. We also spoke with managers responsible for mining units, R&D, and technology acquisition, sales, and service.[5] Many organizations elected to commit more than one representative to this research effort, bringing the total number of individuals participating in the RAND study to more than 90 (see Appendix A).

The organizations and individuals that participated in this study were selected on the basis of their leading positions in the industry and for their ability to

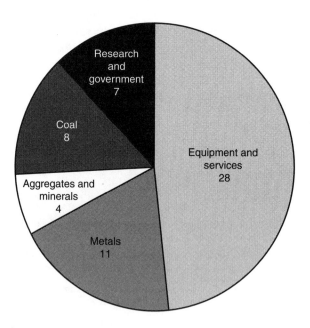

Figure 1.1—Mining Organizations Represented in the RAND Discussions

_____

[5]On several occasions, we were also able to speak with mine foremen and rank-and-file workers.

think broadly and creatively about technology and management issues. RAND selected an initial pool of candidates in close collaboration with the National Mining Association, the National Stone Association, the DOE Office of Industrial Technologies, and industry consultants, as well as labor unions and government representatives. During the discussion phase, participants identified other organizations and individuals as being particularly innovative or having salient technologies and experiences, and in many cases these firms were added to the participant roster.

The industry discussions were conducted between March and July 2000 and consisted of structured, confidential conversations that typically were held on the participants' premises and lasted from 90 minutes to two hours.[6] The discussion protocol, also developed with the input of the aforementioned organizations, is presented in Appendix B.

While we sought to address the specific themes identified in the protocol, we let the discussants determine the issues that were valuable to highlight and those that could be disregarded within the brief meeting time. While keeping the meetings concise facilitated contacting a wide range of executives, it also limited the amount of time we could devote to any one technology prospect.[7] In almost every instance, though, we found the participants to be highly engaged, thoughtful, and willing to address even sensitive issues. On numerous occasions, the host firms generously provided technical literature and facility tours.

Definitions

Technology can be defined broadly as the application of knowledge toward practical ends.[8] Accordingly, technologies include not only physical hardware, but also operational procedures, organizational structures, and management practices. The inclusive nature of this definition is important: According to the industry leaders with whom we spoke, some of the most important innovations at U.S. mine sites concern organization and management.

Mining technology thus includes both the machinery and equipment commonly associated with mining (e.g., drilling, blasting, rock-cutting, loading, and hauling equipment) and technologies that support mining, such as monitoring, control, and communications systems; planning and design tools; and services.

---

[6]Logistical constraints limited our ability to involve a few desired participant firms. To overcome this limitation, three of the 58 discussions were conducted by telephone.

[7]For example, the RAND protocol did not call for quantitative assessments of costs or benefits of critical technologies. Obtaining such information would have been of great value in assessing technology uses and needs but was beyond the scope of the study.

[8]*The Random House College Dictionary* (Rev. Ed.), New York: Random House, 1984.

For the purposes of this study, the definition of *critical technologies* was left to be determined by the participants.[9] That is, the primary attribute that made a technology critical was that it was valued by industry executives. In effect, the critical technologies identified in this study tend to include those that are generally important by virtue of either their widespread application (i.e., the success of mining operations depends upon them) or their distinction as a benchmark (i.e., having a desired feature such as reliability or speed). As a result of this approach, highly specialized technologies (such as those developed to meet a specific environmental objective) tended not to be discussed.

Scope

As noted above, the study participants were selected from the members of the mining community—operating companies, technology suppliers, mining consultants, and researchers. Mining sectors covered in this study include coal, metallic and nonmetallic minerals, and aggregates. The terms *mine, mine site,* and *mining* also refer to quarry operations, except where noted.[10]

The sample should not be considered representative of the entire mining and quarrying sector. In particular, the roster of operating companies is biased toward large and midsize companies.[11] These companies were chosen primarily because they are considered to have valuable experience with advanced technologies—their larger scale and income typically provide greater opportunities to explore technological solutions. The sample of technology suppliers, on the other hand, spans a range of small to large firms. The industrial minerals industry (e.g., phosphorus, potash, clays) is underrepresented in this study, largely due to logistical constraints. However, many of the arguments made regarding the aggregates industry also appear to apply to the industrial minerals industry.[12]

To obtain a broad, high-level view of mining technology trends, we sought out industry representatives in executive- and management-level positions. This approach assured that the discussions considered technology uses and trends within the greater context of the participants' firms, their sectors, and the mining industry as a whole. As a result, the discussions focused more on broad

---

[9]For a discussion of the issue of defining critical technologies, see Steven W. Popper et al., op. cit.

[10]This corresponds with the U.S. Census's North American Industry Classification System Descriptions 212 (Mining—Except Oil and Gas) and 213 (Support Activities for Mining).

[11]The smallest firm in the sample was a gold producer with 125 employees.

[12]This assertion is based on discussions with DOE officials and industry specialists rather than with study participants.

technology trends and often did not entail detailed, "hands-on" knowledge of particular technologies or innovations.

This project focused on technologies applicable to operations conducted at the mine site: engineering development, ore extraction, quarrying, materials handling, and preliminary beneficiation (e.g., crushing, screening, flotation, washing, and concentrating). Prospecting and exploration and downstream processing (such as coal preparation, crystallization, drying, sintering, smelting, and refining) or other activities that typically occur away from the mine site were not addressed.

During the course of the discussions, the industry leaders typically focused on technologies that currently are available on a full-scale commercial or operational basis, as well as those that are poised to be available over the next five to 10 years (i.e., by 2010). This reflects executives' and managers' more immediate time horizons, especially in a very competitive business environment.[13] However, in a few instances, the discussants extended their horizon out to 2020. We asked the industry discussants to exclude ideal or "stretch" technologies and other innovations (even those that have been already developed) if they *are not likely* to be deployed on a production basis in the industry within this time frame.[14]

THE REPORT

To understand the technology development environment in mining, we asked participants to identify trends driving innovation in the industry. In Chapter Two, we discuss some of these drivers, including market forces, regulation, and globalization.

The study participants included a large number of companies working in many different environments and producing a diverse range of goods and services. Nevertheless, when asked to identify leading mine-productivity bottlenecks and the technologies critical to resolving them, participants identified a fairly consistent set of priority areas:

---

[13]In general, participants from academic institutions offered longer-term technology views. In a prior visioning effort, the DOE/National Mining Association Mining Industry of the Future partnership asked industry leaders to identify an ideal development scenario and stretch technology goals out to 2020. The results of that visioning effort, *The Future Begins with Mining*, differed greatly in content and tenor from those of our discussions. See http://www.oit.doe.gov/mining/vision.shtml.

[14]For example, hard-rock fragmentation by means of high-pressure water or plasma jets was mentioned in a favorable light by some participants, although few believed that these technologies would reach commercial production within the next 20 years.

- Unit-operations (unit-ops) capabilities

- Process control and optimization

- Operations and maintenance (O&M)

- Issues relevant to broader organization and management

Within these priority areas, the discussions highlighted dozens of specific technologies that are being employed to enhance mine productivity and are shaping the character of mine and quarry sites in the United States.

The following four chapters address these priority areas and identify many technology solutions being pursued to resolve them. Chapter Three focuses on the basic hardware employed in mining: unit-ops equipment. Chapter Four discusses the application of information technologies (IT) for process control and optimization of mining operations. Chapter Five discusses innovative technologies associated with O&M. A cross-cutting theme present in the discussions was the importance of organization and management to the success of a mine; this issue is addressed in Chapter Six. In Chapter Seven, we close with some broad impressions and observations that we brought away from the discussions.

Each chapter presents the status of the technologies, trends in diffusion, and distinctions in application among mining-industry segments. Examples of applications are highlighted in boxes. In line with the project's goal of promoting frank discussion, we do not attribute comments or technologies to specific individuals or companies.

We emphasize that the slate of critical technologies identified in this report is not intended to be exhaustive. Rather, the report summarizes and synthesizes the perspectives of a sample of industry representatives, which are based on their understanding of their firms' mining-related activities and the industry at large. The absence of a particular technology in this analysis should not be interpreted as an intentional omission, but as an indication that the technology was not cited as critical during the discussions in this study.

We also note that this report does not present a consensus view. While we discovered many points of convergence during the 58 discussion sessions, we also heard many conflicting arguments. While some discussants spoke enthusiastically about the prospects for a new technology, others squarely disparaged it. We flag converging and conflicting views to illustrate the breadth of thought and experience across this very diverse industry.

Chapter Two

DRIVERS OF AND IMPEDIMENTS TO TECHNOLOGY CHANGE

To establish the context of technology development and diffusion at U.S. mine and quarry sites, RAND asked participants to address the process of technology innovation in the industry. In other words, what factors determine the development and diffusion of new technologies at U.S. mine sites? And how will technology flows evolve in the coming years? Several drivers of and impediments to innovation were discussed, including

- Industry demand for new services and equipment

- Commodities markets

- R&D funding and alliances

- Regulatory and community constraints

- Industry consolidation

- Globalization

In view of these trends, industry representatives suggested that technology innovation in the mining industry for the foreseeable future was likely to be characterized by slow but steady incremental improvements. Innovations that require significant private sector R&D expenditures, capital investment outlays, and facility reengineering are likely to be viewed as too speculative and costly in the current industry climate.

INDUSTRY DEMAND FOR NEW SERVICES AND EQUIPMENT

The United States has the largest mining industry in the world, with a raw material production value of over $52 billion in 1997 (see Table 2.1). Yet many industry representatives noted that as a buyer of goods and services, mining is relatively small in comparison with other industries, and its ability to finance R&D specific to mining is limited. As a result, many technology innovations in mining are adopted from other sectors such as construction, automobiles, and

Table 2.1

The U.S. Mining and Quarrying Industry, 1997

Item	Total	Coal	Metals	Nonmetallic Minerals
Value of shipments ($ billions)	52.3	24.0	11.5	16.8
Employment	239,400	93,000	48,500	97,900
Number of establishments	7,350	1,511	493	5,344

Source: U.S. Census Bureau.

aerospace. An example of technology crossover is the Global Positioning System (GPS).

Although the mining sectors in Australia and Canada are smaller than those in the United States, their local economic impacts are greater, and mining in these countries has received significantly more public- and private-sector R&D support (see Figure 2.1). Not surprisingly, many discussants pointed to innovative mining technologies coming out of these countries that may have an impact in the United States.

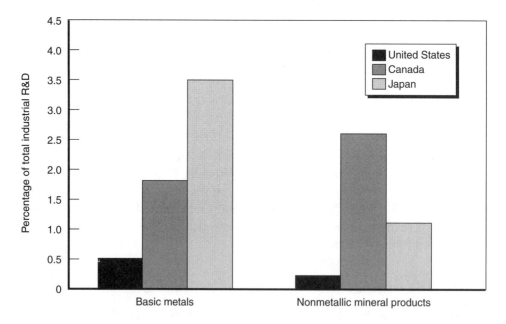

Source: National Science Board, *Science and Engineering Indicators 2000*, NSB-00-1, National Science Foundation, 2000.

Figure 2.1—Mining-Related R&D as a Percentage of Total Industrial R&D, 1996–1997

According to many discussion participants, the mining industry as a whole tends to be risk-averse in its application of technology. Several speakers repeated an industry saying: "Miners like to be first to be second." "The inertia to change is large," said a service provider. Many causes for this were cited, including the basic nature of the raw-materials production process, volatile commodities markets, and, interestingly, industry culture. Operating-company representatives frequently pointed to other enterprises or units as being technology leaders in one aspect or another, but they typically painted their own organization as more conventional.

Economic factors tend to favor risk-averse technology decisions in mining. The volatility and uncertainty inherent in commodities markets raise the perceived cost of long-term planning and investment. Because of the large scale and complexity of many pieces of earthmoving equipment and the need to coordinate technology acquisitions with mine-development plans, investing in new technologies often entails massive capital investments. Yet the period required to recover the cost of investments in mining tends to be much longer than in other industries, because of the highly competitive nature of the business and resulting thin margins.

> Firms in the construction industry typically expect to obtain a return on their technology investments within three to six months. In the mining industry, investments are often not recouped for two to three years.
>
> —*Industry representative*

The mining industry purchases relatively few pieces of equipment. Moreover, because of its scale and ruggedness, mining equipment often has no application in other sectors. Thus, R&D, demonstration, and start-up production costs per number of units sold are high. As an example of one extreme, the most recent sale of a new dragline for surface mining in the United States was made in 1993. The time required to develop and recover the development costs for large-truck tires is currently 10 years, reported one industry representative, and this has constrained the scale-up of haul trucks. With low turnover, the capability to prototype or make design improvements based on field experience is limited. As equipment increases in size, power, versatility, and durability, mining companies are reducing the number of units they need to purchase, thereby extending the development-cost recovery cycle.

REGULATORY AND STAKEHOLDER PRESSURES

During the course of the RAND discussions, federal health, safety, and environmental regulations were rarely cited as a major driver or inhibitor of technology innovation in mining. However, when specific regulatory issues were raised that have been under discussion at the federal level (concerning, for example, diesel emissions, air quality, ergonomics, and noise), industry representatives tended to argue along two lines: On the one hand, speakers often implied that potential regulations could be met as part of the normal technology development and diffusion process. New standards to reduce diesel emissions, for example, were not seen as a technology concern by any mining company or equipment supplier. Rather, it was argued, the diesel regulations somehow would be addressed by engine manufacturers. "We tell them what we need, they do all the work," said one mining-equipment representative.

On the other hand, discussants argued that application of existing regulations and prospective regulatory changes that are seen as impractical (proposed coal-dust and noise regulations were cast in this light on several occasions) simply would result in a halt to mining operations. Speaking of the lengthening environmental permit process for new mines, a coal-industry executive described the situation as "an effort to discontinue what we are doing entirely." In other words, industry representatives believe that such regulations have passed beyond the realm of technology solutions.

Complying with state and local regulatory efforts—concerning land-use permitting, for example—also was seen as impractical from a technology perspective. While stone and gravel producers currently enjoy strong demand for their products, they also face increasing demands to regulate hours of operation, noise, visual impacts, road use, and groundwater impacts, especially at interior (in-town) quarries. Yet there are few economically feasible technology solutions for complying with such regulations in real time. Reducing the footprint of a quarry by, for example, reducing on-hand product inventories and outsourcing support services was cited as a potential solution for some sites. In Montana and Colorado, ballot initiatives to ban the use of cyanide in gold production have been mounted (the Montana initiative was successful, the Colorado initiative was not). Several industry discussants pointed out that no economically feasible alternatives to cyanide are currently available, but three executives from technology supply firms did suggest that in the event of a ban, economic imperatives would drive a shift to alternative leaching agents.

The development of new mines and quarries in the United States has been constrained in recent years by a combination of factors: economic pressures, environmental and land-use regulations, and political constraints. As a result, technology innovation in U.S. mining is devoted largely to reengineering processes and facilities to boost productivity, increase recoverable reserves, and extend the life of existing facilities. One outcome is that some operations are mining much deeper than originally envisioned, resulting in greater technological demands being placed on logistics, utilities, and safety systems. Deeper operations combined with environmental and aesthetic concerns may motivate more open-pit mines and quarries to shift to underground operations in the future—what two representatives described as "pulling a cover over our head."[1]

> In the coming years, land-use restrictions are likely to result in a drop in the number of quarries in America. Operations that remain open will be located farther away from populated areas and will have a larger average size.
>
> *—Aggregates-industry representative*

The prevalence of mature mining operations in the United States tends to constrain innovation because of a hesitance to abandon legacy systems (e.g., existing machinery and mine layouts) and engage in costly reengineering exercises. In contrast, discussants often pointed to newer mines in South America, Africa, and Australia as being technology leaders in areas such as planning and automation.

Regulation rarely was cited as an inhibitor of innovation. However, on one occasion, a speaker observed that roof-support technologies were advancing quickly, but that the existing regulatory framework (i.e., the lack of guidelines for application of new technologies) has slowed their diffusion. And another executive noted regulators' reluctance to abandon stench gas in favor of newer hazard-warning technologies perceived to be more effective.

Rather than speaking about government mandates, discussants pointed to their voluntary initiatives. Corporate policies that place a primacy on health, safety, and environment have become widespread in the industry. Improvements in worker safety and health frequently were presented as a function of good management and the need to hire, retain, protect, and motivate both management and staff employees in a competitive and selective labor market. Similarly, innovations in environmental protection were portrayed as a function of proactive risk management, good public citizenship, and maintaining employee morale.

COMMODITIES MARKETS

Over the past several years, prices for many mineral commodities, including gold, copper, and coal, have been at or near long-term market lows, thereby

---

[1]There are about 5,300 stone and aggregates quarries in the United States. According to one source, the United States has about 100 underground quarries in operation, with another 25 in planning and development. A stone and aggregates producer with whom we spoke estimated that his company's underground operations may double, increasing from six to 12 by 2020.

shrinking margins across much of the industry (see Figure 2.2). This has caused mining activity for metals in particular to decline (see Figure 2.3).

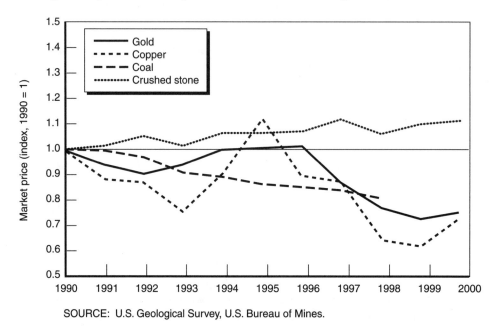

SOURCE: U.S. Geological Survey, U.S. Bureau of Mines.

Figure 2.2—Market Prices for Copper, Gold, Coal, and Crushed Stone

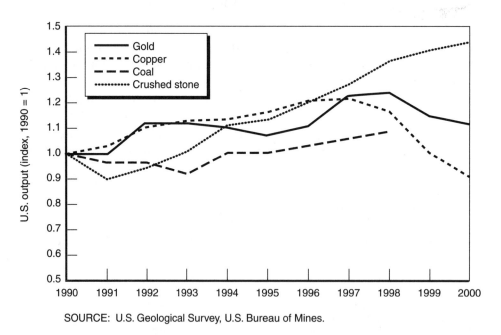

SOURCE: U.S. Geological Survey, U.S. Bureau of Mines.

Figure 2.3—U.S. Output of Copper, Gold, Coal, and Crushed Stone

Interestingly, industry executives presented two strikingly different views of the technology implications of a difficult business environment. One set of executives argued that low market prices for their products and the resulting thin operating margins impair their ability to raise the capital (from both operating revenues and capital markets) needed to invest in new machinery and equipment and thus constrain innovation. Indeed, many coal and metals companies have been forced in the last few years to defer spending, idle and mothball operations, and restructure their enterprises—"in conservation mode," as described by a gold-company executive. A historically large number of distressed operations have been put up for sale. In this environment, executives are particularly wary of new and unproven technologies and technologies that may require major reengineering of mine operations.

> "The year of 1999 was a difficult year for [the company] due to the lowest gold prices of the last 20 years. . . . The focus of [the company's] activities during the year continued to be to reduce debt and operating and corporate costs, improve liquidity, and enhance and preserve the company's important gold assets."
>
> *—Gold-company annual report*

> "Anticipating reduced near-term market demand, [the company] curtailed its capital investment plans in 1999 and expects to further reduce its capital spending in 2000. Current efforts are directed at optimizing production from existing mines with the lowest production costs."
>
> *—Coal-company annual report*

Another set of executives explicitly refuted this perspective, arguing that the prevailing harsh financial climate in metals and coal is motivating managers to make major changes in their operations in efforts to achieve significant productivity breakthroughs. A metals-company executive reported that his firm had "significantly changed the way we work" at its U.S. facilities by reengineering the facilities and bringing in new equipment such as jumbos[2] and remote-controlled vehicles. He added that these technology investments represented "a huge change" in the company's approach to mining. Several discussants argued that their decision to invest in new technologies in the current period was based on a strategic calculation that commodity prices might not rebound in the foreseeable future.

---

[2]A jumbo is a drilling apparatus used for drift development in underground metal mines.

> "[The] ability to generate cash throughout the year, coupled with our strong balance sheet, enabled us to actively acquire aggregates companies and fund an aggressive capital investment and expansion program within our heritage businesses."
>
> —*Aggregates-company publication*

An important exception in the sector is the stone and gravel industry, which has been enjoying historically strong demand and growing revenue streams as a result of the record-long U.S. economic expansion and generous public-sector spending on infrastructure (see Figures 2.2 and 2.3). Many stone and gravel producers are using their cash to acquire weaker firms and invest in new technologies to quickly ramp up production and capitalize on a favorable market environment.[3]

RESEARCH AND DEVELOPMENT FUNDING AND ALLIANCES

According to industry executives, cuts in R&D by government and industry are likely to result in fewer fundamental or breakthrough technology innovations in the future.

Over the past three decades, many participants noted, mining concerns in the United States have scaled down or eliminated entirely their R&D operations—a function of trimmed profit margins and a broader business trend of focusing on "core competencies." This decrease is reflected in the low rankings of mining-related industries in a comparison of R&D expenditures across industry sectors (Figure 2.4). In addition, dramatic cutbacks in federal funding for industry since 1988 have reduced budgets for advanced R&D in both academia and the private sector.[4] As a result, almost all mining companies said that their mining-related R&D activities (if they reported having any) were largely confined to short-term and site-specific problem-solving. This has shifted the locus of technology research, development, and demonstration to technology providers.

As mining technologies become more complex and mining processes become more tightly integrated, the need for sustained, strategic alliances between equipment developers and mine operators is becoming more critical. Few organizations have the capability to combine metallurgy with machine design to

---

[3]Industry discussants suggested that the strong and steady market for most industrial minerals, such as aggregates, is likely to support continued technology investment in that industry as well.

[4]National Science Board, *Science and Engineering Indicators 2000*, NSB-00-1, National Science Foundation, 2000. The principal source of federal funding for mining technology R&D was the U.S. Bureau of Mines, which was abolished in the mid-1990s. The discussants, however, were divided on the practical technology implications of this act.

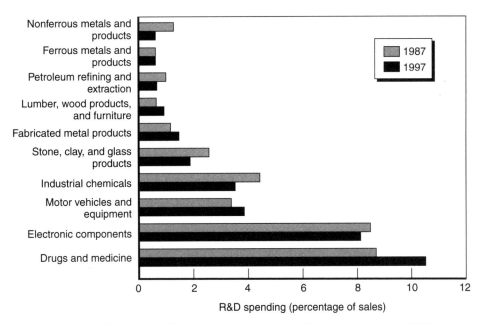

SOURCE: National Science Board, *Science and Engineering Indicators 2000*, Arlington, VA: National Science Foundation, 2000 (NSB-00-1).

Figure 2.4—U.S. Nonfederal Industrial R&D Spending as a Percentage of Sales

develop advanced rock-cutting technologies, one industry executive noted. Similarly, the development of automated equipment requires coordination and collaboration among producers of machinery, communications and GPS, sensors and imaging technologies, and control algorithms. But funding from both the private and public sectors to catalyze and sustain such partnerships has been very limited in recent years.

Some operating firms have begun to reconsider the trend away from investment in R&D by searching more widely for new technologies to apply and by expanding their in-house R&D activities. For some, the driver was the desire to achieve productivity breakthroughs in a difficult business climate.

• A stone and aggregates producer has convened an in-house "technologies task force" to conduct a systematic survey of other industries, as well as government laboratories, for technologies to apply to its quarry operations.

• Reconsidering its strategy dating back to the 1970s, a metals producer has rebuilt its R&D center over the past three years. The staff of 40 has undertaken more than 200 research initiatives—including more "far-out" speculative ventures in remote sensing and materials science.

Innovation often springs from insights gained through the technology buyer/supplier relationship. "You need everyone at the table to work things out," said one manager about the ideal innovation environment. Yet one technology developer characterized the current situation as a "stalemate." According to several discussants, many operating companies are not particularly interested in alliances, risk-sharing, and pilot-testing new technologies. Several mining company representatives stated that they wanted to use only those technologies that already were proven on a commercial basis. Said one supplier, "Not too many of them want to be first at anything." As if in response, another technology provider quipped, "We never send a new product out the door with 'Serial Number One' on it."

INDUSTRY CONSOLIDATION

In response to commodity-price pressures and the desire to achieve economies of scale, the mining industry is undergoing consolidation, among both operating firms and technology suppliers. Most participants expressed the expectation that this consolidation would continue or even accelerate over the next decade. Consolidation presently is having an especially strong impact on the profile of the stone and aggregates industry in the United States, which counts an estimated 5,000 operations and traditionally has been dominated by small-capitalization "mom-and-pop" organizations. Meanwhile, the coal and metals segments are continuing a trend of consolidation that has been under way for several decades.

Consolidation is likely to have an impact on facility size and age. As operating firms consolidate, the number of mines is likely to fall, and the average size of the remaining operations will increase. With fewer small and medium-size firms, the merits of developing smaller ore deposits and shorter-lived facilities (i.e., niche operations) will diminish.

Consolidation could, in theory, have a positive impact on the flow of new technologies in mining. Consolidation among technology suppliers should result in economies of scale in R&D (especially as mining equipment becomes more complex and costly) and should encourage greater technology integration, for example, of IT and unit-ops. Larger firms also tend to have greater financial resources and know-how to identify, evaluate, and acquire new technologies, and mergers and acquisitions offer important opportunities for units to share their best practices and technologies and bring operations up to an even level of technical proficiency.

"A key benefit of unification is the opportunity it gives us to share, determine, and benchmark best management practice in every area of our business—from customer service to employee development and operating standards."

—Aggregates-company publication

However, several technology developers expressed concern that consolidation could retard technology innovation, arguing that as mining companies grow in market share, they also gain in their ability to drive down equipment prices and to push more R&D costs back onto suppliers. In February 2000, 14 mining companies announced the creation of an internet-based mining exchange to centralize and streamline acquisitions. Several technology providers openly wondered if the move to an on-line machinery and equipment market would further erode suppliers' margins, preclude value-adding opportunities, and rend buyer-supplier relationships that have led to past technology innovation.

GLOBALIZATION

As the industry consolidates, mining companies and technology providers are becoming increasingly global. Economic liberalization and falling barriers to international trade also are encouraging firms to invest in East Asia, Africa, South America, Eastern Europe, and the former Soviet Union, and mines in these countries are now important customers for U.S.-based technology developers. Meanwhile, many foreign mining companies and equipment and service providers have invested heavily and acquired operations in the United States.

Executives with whom we talked had diverging opinions about how globalization will affect the mix and pace of technology innovation at mine and quarry units in America: Some implied that it had a neutral or restraining effect, but most argued that globalization tended to favor technology innovation and diffusion in the United States.

One major underground-mining-technology provider reported that the equipment his company provides to operations abroad often is one or two generations behind the latest technology. Representatives at a U.S.-based multinational mining company pointed out that logistical constraints often limit the ability to implement cutting-edge practices in remote and poorly serviced locations abroad. An aggregates-industry representative pointed out that while lessons learned from Western European practices have prompted higher standards for road-building materials in the United States (and hence the demand for improved crushing and screening technologies), U.S. quarry operations are, on average, more innovative, due to the higher demand for construction materials and greater competition within the local industry.

Most mining-company representatives argued that their operations abroad were at a level of technology application comparable to that of their units in the United States. They noted that key staff members are rotated throughout a global company, and they pointed to the cost advantages of standardization and groupwide purchasing across a firm's operations. However, three technology suppliers argued that groupwide sourcing can have a negative impact on technology innovation: By seeking a single supplier and common technologies for all of their operations, regardless of local conditions, operating companies can overlook technologies that may be more innovative.

Several argued that with limited R&D and few new mining developments in the United States, innovations from operations in countries such as Australia, Indonesia, Peru, and South Africa will play an increasingly important role in influencing the character and management of mine sites in the United States. For example, advanced mine modeling and design software developed and widely deployed in Australia is now being adopted in the United States. The first U.S. application of an automated ore-stacking technology developed by a U.S. company came only after several successful installations abroad.

Chapter Three

CRITICAL TECHNOLOGIES FOR UNIT OPERATIONS

Mining typically is conducted as a series of discrete steps, or unit-ops, which we examine in this chapter:

- Drilling, blasting, cutting, and excavating

- Ground control

- Loading and hauling

- Materials processing

Given their centrality to the mining process, most unit-ops machinery and equipment were cited as critical technologies by participants. However, industry representatives tended to devote attention primarily to the latter phases: materials loading, hauling, and processing. Drilling and blasting technologies were rarely cited by operating-company representatives as critical.

One of the principal goals guiding unit-ops technology development is that of reducing the cyclical or batch character of mining, making a mine operate more like a refinery. With few exceptions, unit-ops technologies in use today at mine sites in the United States will be in use in 2020. Technology changes that are being implemented are incremental and typically focus on increasing batch size, reducing cycle intervals, and boosting equipment availability. Expected commercial introductions for several unit-ops technologies are summarized in Table 3.1.

DRILLING, BLASTING, CUTTING, AND EXCAVATING

According to technology suppliers, drilling- and blasting-related technologies are rapidly evolving and are offering mining companies important new capabilities. Nevertheless, operating-company representatives rarely identified drilling and blasting technologies as critical during the RAND discussions. For

Table 3.1

Anticipated Availability of Selected Unit-Ops Innovations

Technology	Anticipated Date
Solid-state programmable blast detonators	2000
Six-unit miner-bolters	2000
Mechanical cutter for hard-rock applications	2003
Fuel-cell-powered underground equipment	2010
1000-ton-capacity haul truck	2020
150-cubic-yard-capacity shovel	2020

Source: RAND discussion participants.

one supplier in particular, this disparity in perceptions appeared to preclude important productivity-enhancing opportunities.

The most significant changes in blast technology address product-delivery systems. One factor is the continuing trend away from the use of cartridged products in favor of bulk products for both surface and underground operations. Representatives mentioned new surface and underground delivery-vehicle technologies that boost blast accuracy and safety: high-precision pumps and blending and measurement devices, robotic arms that place the product in the hole, and remote controls. The integration of IT is also improving blast optimization capabilities through in-field measurement and reporting of loading information and blast results such as particle size, heave, and distribution. (This subject is discussed further in Chapter Four.) A third trend is outsourcing of blasting-related services, ranging from consulting on safety to providing comprehensive packages priced according to the volume of "shot rock" on the ground or ore processed. As a result of these blast-optimization efforts, the volume of blast agents per unit of shot rock is slowly decreasing.

The standard blasting agent throughout the industry remains ammonium-nitrate-based emulsions, and this basic explosives chemistry is unlikely to change in the coming two decades.[1] While alternative blasting agents (e.g., water gels) are available, they are rarely used in mining because of their higher cost (three to four times the cost of conventional emulsions). When considering blasting technologies, operating companies tend to be highly cost-conscious, which mitigates opportunities to develop value-added or innovative products. Two explosives representatives noted that quarry operations, though a smaller share of their business than coal and metals operations, offer greater

---

[1]One supplier noted that ammonium-nitrate emulsions accounted for 90 percent of their explosives sales in the United States.

opportunities for higher-margin business relationships due to operators' desire to expand capacity.

> Explosives are a mature technology and have become a commodity product, motivating suppliers to look at the entire blasting-related value chain for opportunities to capitalize on their expertise. As a result, providers are shifting their business focus from products to service.
>
> • One technology supplier has signed a covenant to pursue "special projects" with a long-time customer to improve the latter's coal-production processes. The partners have agreed to share the profits. This arrangement resulted from "a paradigm shift" in the coal company's view of the value of outside expertise.
>
> • Arguing that blasting generally is conducted according to rules-of-thumb dating back to the 1950s, one supplier noted his firm's ability to draw on its database of over 20,000 blast configurations to help mining companies optimize their blasting and reduce their haulage and processing costs.

A small number of participants cited solid-state programmable blast detonators as an important technology that is currently coming to market. Advantages include more-precise delay timing (resulting in increased blast efficiency and control) and greater compatibility with remote-controlled loading of explosives and wireless detonation. However, these initiating systems have significantly higher costs, and none of the participants reported routinely using them. A future niche market may be in-town quarries under pressure to reduce off-site vibration and noise impacts.

Several operating companies and technology and service providers expressed concern about nitrogen-oxide (NO_x) releases from blast sites and their potential health impacts on workers, as well as their aesthetic and environmental impacts on nearby communities.[2] These concerns may increase as bulk mining of metals and cast-blasting in surface coal operations become more prevalent. But one blasting specialist noted that NO_x releases are rare and that there has been no reliable quantification of occupational or ambient impacts.

After many years, continuous miners and longwall systems remain the principal underground coal-extraction technologies in the United States, the output from both methods accounting for more than 90 percent of all underground coal produced. Output from longwall coal-production systems in the United States surpassed output from continuous mining systems in 1994, and, given the con-

---

[2]NO_x is released as a result of incomplete combustion due to the lack of proper confinement or the incursion of water into the blasting agent. These problems are being addressed through advances in blast modeling and design. An operating-company executive noted that residual nitrates also have the potential to contaminate groundwater.

tinuing growth of longwall panel productivity, the share of underground coal output produced by longwalls is likely to continue to grow for the foreseeable future.

A few incremental enhancements were mentioned by study participants. Longwall panel dimensions increased substantially in the United States during the 1990s: Average face widths increased from 600 to 1,000 feet, and panel lengths grew from 5,000 to 7,500 feet. State-of-the-art systems currently have face widths of over 1,200 feet. According to one coal-company representative, extension of longwall widths is limited primarily by the capabilities of the face conveyor to pull coal over longer distances: Space constraints limit the size of drive units, and existing materials and engineering capabilities limit the strength of drive chains. Additional increases in drive output will require supplying higher voltages at the face (above the current 4,200-volt peak standard) and may present a greater safety hazard, namely, fire and electrocution. Other technology constraints to further boosting longwall productivity include horizon control, ground control, and ventilation.

During the RAND discussions, rock-cutting technologies were identified as a promising area for innovation.

• A member of the research community noted that an "oscillating disk cutter" was in the early stages of development and may be available to the mining industry by 2010. Another researcher, however, disparaged prototype demonstration efforts and claimed that the basic concept—in which the cutter strikes the rock—was unworkable.

• A technology supplier reported that a continuous mechanical cutter his firm was developing was "on the edge of breaking through the [feasibility] envelope" and would be commercially available by 2003. The technology, which entails specially hardened conical picks laced on a cutter head, was most likely to see its first U.S. production applications in the Carlin Gold Belt of Nevada.

• Tunnel-boring machines are an alternative, albeit more costly, rock-cutting technology occasionally used in mining. An underground metal operation reported that it was using a commercial tunnel-boring machine to develop a major drift as the mine transitions from surface to underground operations.

• An aggregates-company executive noted that his firm was seeking to develop and deploy machinery that cuts hard rock in a fashion similar to that of pipe-trenching and marine dredging equipment.

Several discussants raised the issue of mechanical cutting of hard rock as an alternative technology to blasting.[3] As has been experienced with coal and other

---

[3]Mechanical cutting technologies for soft-rock materials such as phosphates and trona have been commercially available for several decades.

soft minerals, mechanical cutting of hard rock would play a critical role in eliminating the batch processing character of drilling and blasting and would also produce more uniform particle size. In addition, reduced shattering and shock damage to surrounding material would reduce ground-control requirements in underground operations. The need to find an alternative to blasting to reduce percussive noise at in-town locations is driving a search for rock-cutting technologies for the quarry industry. The key constraint in developing such technologies, an executive noted, is the problem of raising the compressive strength of head wear materials enough to cut consistently and economically through the range of rock found in a mine. Other technology challenges are the mitigation of noise and dust generation.

GROUND CONTROL

Ground control was most commonly cited as the unit operation undergoing the most important technological innovation in underground environments.

Miner-Bolters

The majority of ground-control operations in underground mine development are accomplished through rock-bolting. Accordingly, the introduction of a new generation of combination continuous-miner/bolters was raised as an important innovation. Standard commercial miner-bolters consist of a continuous miner equipped with two bolting units and require three to four operators. New-generation miner-bolters have six bolters (four roof units and two wall units) that operate while the continuous-miner cutting barrel is advancing. These new miner-bolters require as few as two operators. They are presently available from one European manufacturer, and a U.S. manufacturer is scheduled to release a model in late 2000.

New-generation miner-bolters are likely to have a major influence on both the method by which coal-mine entries are developed in the future and the speed of those developments, according to industry representatives. Because the new miner-bolters complete all the required bolting on the first pass of the equipment, no time is wasted cycling machinery, support equipment, and supplies in and out of entries.[4] In principle, the machines do not need to stop until the entry is complete. In practice, however, other constraints (e.g., coal removal, ventilation, bolt and resin supply, machine reliability) may become limitations at some point.

---

[4]Presently, entry development typically proceeds by place-changing (or "cut and flit"): A continuous miner (with no bolting capability) advances 20 to 40 feet and then must retreat from the entry (at which point it begins cutting in a different area) and be replaced by a bolting team.

Cable Bolts

A second important ground-control innovation is cable-bolting for primary support. Cable bolts—constructed from multistrand steel cable—have been an important ground-control technology for several years, primarily as a form of supplemental roof support.[5] Industry participants noted that cable bolts, because of their significantly higher tensile strength and resistance to shear failure relative to solid rod bolts, permit greater bolt spacing; they also reduce—and in some cases may eliminate—the need for supplemental support. This reduces material costs and development times and improves productivity by minimizing the jockeying of equipment. Indeed, the productivity benefit realized in the transition to cable bolts may be on a scale comparable to that of the transition from timber to bolting, said one discussant. Cable bolts will become increasingly attractive as mines continue to go deeper and supplemental support requirements increase.

Primary cable-bolting presently is employed in one longwall coal mine. Because the surface area of cable bolts is much higher than that of solid rod bolts, general-use approval of cable bolts is contingent upon demonstrating long-term corrosion resistance. The latest designs, which incorporate epoxy coating and galvanization, may provide adequate resistance. Some participants claimed that the use of cable-bolting also may facilitate remote-control or semiautonomous bolting when bulk bolt material is stored on a spool.

Spray Coatings

A third trend in ground control, primarily in metals mining, is the increasing use of spray coatings in primary-support applications.

Concrete spray coating (shotcrete or gunnite) traditionally has been used for roof-support rehabilitation and repair, but its application as a primary roof support may increase due to the availability of new sprays featuring advanced concretes, polymers, epoxies, and fiber reinforcements.

According to manufacturers with whom we spoke, these coatings are not necessarily meant to replace bolting; rather, they are intended to provide more flexibility in meeting roof-support needs. Shotcrete, for example, may be used in conjunction with bolts to eliminate screening, permit wider bolt spacing, and decrease bolt length. Polymer or composite coatings, while contributing little structural support, provide a watertight seal on the rock surface, reducing the roof's susceptibility to chemical weathering and degradation.

---

[5]Supplemental support entails additional bolting in a previously supported area. This may be required as a result of excavation elsewhere in the mine altering the local stress field.

Synthetic reinforcement fibers for spray coatings. Fibers are one of a number of new technologies that enhance the strength and durability of shotcrete, enable its use as a primary roof support, and reduce snag injuries to personnel.

Suppliers noted that an advantage of spray coatings is the ability to install support sooner after blasting than is possible with bolts. In addition to decreasing the production cycle time, rapid application apparently preempts the critical stage of initial roof sagging and "fools the rock" into needing less support. Another cited advantage of spray coatings is the ease with which they can be applied via remote operation, thereby eliminating the need to have an operator working under an unsupported roof.

Manufacturers and mining-company representatives indicated that the use of primary shotcrete in underground metal mines is increasing. But while manufacturers have made a strong case for the benefits of spray coatings for primary support, this technology is still used in only a small fraction of operations. The widespread use of polymer coatings has been limited because of concerns about health and fire hazards. These problems may be circumvented, however, with the recent introduction of a new generation of composite concrete-polymer materials.

LOADING AND HAULING

Haul trucks, shovels, and excavators are typically cited as the most critical technologies for surface-mining operations and are the units around which most mining operations are designed and planned. The most significant technology innovations, from the perspective of the executives with whom we spoke, are incremental and address load capacity and reliability.

In the late 1990s, four haul-truck manufacturers broke through the 300-ton barrier, and truck manufacturers are expected to increase the capacity of trucks even further in the future. "Super-size" 400-ton trucks are currently in testing,[6] and one supplier predicted that by 2020, 1,000-ton-capacity haul trucks will be commercially available.[7] As trucks scale up over the coming years, so will excavators—from the current 50-cubic-yard rating to 150 cubic yards.

A 50-cubic-yard electric shovel loads a 240-ton haul truck in three passes. The scale-up of excavators is expected to follow in tandem with the scale-up of haul trucks.

---

[6]One coal producer reported having three 400-ton trucks in operation.

[7]One participant claimed that the latest increases in haul-truck ratings are misleading: Some models rated at 340, 360, and 400 tons, for instance, are of the same basic technology as 320-ton trucks. Manufacturers simply upgraded the rating in response to *de facto* usage patterns of mining companies.

At quarry operations, the standard haul-truck capacity has increased from 35 to 50 tons over the past several decades. Now, 85- and 100-ton trucks are in use at some facilities in this segment, said a discussant. This capacity-growth trend is likely to continue as older, smaller, in-town quarries are closed in favor of larger, more remote operations.

Few major changes were reported in underground haulage systems. The most critical change with load-haul-dump (LHD) vehicles, said one executive, is the horsepower-to-weight ratio, which is boosting tramming speed on grades. This has particular salience for mines that are converting from surface to underground operations and that use inclined drifts. A continuing trend away from tethered (electrical) underground equipment was expected by one discussant. Two discussants raised the merits of alternatives to diesel equipment, e.g., hybrid diesel-electric battery systems and natural-gas engines. Fuel-cell technologies were seen by five discussants as encouraging, and one equipment manufacturer asserted that fuel-cell-powered underground equipment would be available by 2010.

The study participants highlighted performance enhancements and constraints associated with critical subsystems. One participant drew attention to the importance of faster-cycling hydraulic systems: Hydraulic systems are being reengineered (e.g., by reducing the number of cylinders) to reduce the cycle time of loaders, excavators, and haul trucks.

Truck tires are a critical subsystem technology influencing the development of heavy haul trucks.

• Several participants cited tire load and speed ratings as the main factors (functions of tire materials and construction) limiting the scale-up of haul-truck capacity. Also, as tires increase in size, they impinge on the ability of truck manufacturers to build structurally sound frames.

• Tires were cited by one executive as the principal productivity bottleneck for his firm's mining operations.

• Other tire-related concerns raised by discussants were cost and disposal. A tire manufacturer reportedly is developing a three-piece modular tire design that will reduce manufacturing and maintenance costs.

A prominent debate that emerged during the discussions concerned the appropriate scale of loading and hauling equipment, especially for surface mining. Reviewing the rapid increase in truck and excavator capacity over the past several decades, some participants argued that increases in equipment scale, when appropriately applied, tend to boost productivity. Several mining and quarrying executives and technology providers looked forward to the continuing scale-up of haul trucks, loaders, and excavators and the expected economies this scale-up would bring.

However, some participants questioned whether the size of haul trucks and excavators has reached a feasibility threshold where the economies of scale have peaked. They raised several issues:

- Haul-truck size has physical limitations, including space constraints on the further scale-up of drivetrains and the need to maintain structural integrity of equipment bodies.

- Increased equipment size results in greater production and transportation costs, greater numbers of pieces that must be shipped to the site, and longer field assembly times.

- Scaled-up equipment requires extensive facility reengineering: Bigger trucks require heavier-duty haul roads and bigger maintenance sheds and must be paired with larger-capacity excavators.

- There appear to be no near-term breakthroughs in tire load capacity.

- A *de facto* size limit is imposed by the number of shovels in operation: Most facility managers would like to keep at least three shovels operating, for enhanced flexibility and reliability.

One participant cited a need to focus on alternative ways of increasing truck capacity and efficiency, such as developing lighter-weight truck bodies, higher-capacity wheel bearings, and higher-volume truck beds.

> Shovel sizes are nearing the point at which the roller circles on which the machine turns can no longer be shipped easily in one piece. Beyond the added shipping demands, field assembly of this part involves welding and subsequent machining, an expensive and time-consuming step that requires specialized tooling and personnel on-site.
>
> *—Equipment manufacturer*

Where feasible, underground and surface mines are replacing track and truck haulage with belt-haul conveyors, which have a lower operating cost (see Table 3.2). Coal companies in particular highlighted the increasing use and importance of conveyors. The conversion to belt haulage systems was listed as one of the top three current mine-site investment priorities of one major coal producer. Another coal producer argued that running belts through mined-out tunnels was more flexible and cost-effective than building and maintaining sur-

Table 3.2

Belt and Truck Haulage Costs

Method	Cost (cents per ton/mile)
Truck haulage	15–35
Conventional belt haulage	5–10
Automated belt haul and stacking	1–15

Source: Industry representative.

face roads, and that it reduced labor costs in the mine. He described his company's operations as now having "a tremendous amount of belt structure."

MATERIALS PROCESSING[8]

One major metals producer cited ore crushers as an important weak link in his company's operations. Anomalous incidents, such as the accidental introduction of a shovel bucket tooth, were cited as a major issue with which he had to contend. Crusher reliability is especially critical, because an interruption in downstream operations can be very costly.

Many aggregates producers are deploying mobile crushers and screeners to ramp up production to meet steadily growing demand stimulated by rapid growth in federal and state transportation spending as well as the strong demand in the construction industry. Other goals are to manage production capabilities among a firms' various quarry operations more flexibly and to provide local dedicated capacity on-site at large-infrastructure projects. Although they add flexibility, mobile crushers have relatively short working lives and require maintenance overhauls after seven to 10 years, according to one representative.

Vertical impactors are gradually replacing cone crushers in the aggregates industry to boost throughput and to allow operators to meet increasingly tight buyer specifications that call for aggregates of a consistent "clean" grade and more precise shape.[9] However, the operating temperatures and throw rates of vertical impactors are higher than those of cone crushers. Vertical impactors

---

[8]This study focused on materials-extraction activities at the mine site. But in some discussions, most notably those with aggregates producers, materials processing was seen as an important mine-site activity.

[9]Road-building standards and contracts in the United States increasingly call for the use of Superpave, a high-performance asphalt that uses cubical rather than conventional oblong-shaped aggregate.

also tend to have shorter wear lives (five to 10 years) and require more-frequent overhauls.

Several discussants noted the trend toward greater use of leaching processes for winning metals from oxide as well as sulfide ores. Two drivers were identified: the decreasing availability of high-grade ore in the United States, and the growing capabilities for bulk ore handling and processing. One technology highlighted by a discussant is a reusable leach pad serviced by an automated "race track" ore-stacking, leaching, and bucket-wheel harvesting system. According to the technology provider, the automated system boosts ore recovery rates from approximately 30 percent to between 85 and 90 percent while significantly reducing energy and labor costs. The continuous, conveyor-fed process also reduces costs associated with conventional batch (truck) haulage and stacking.

Chapter Four

CRITICAL TECHNOLOGIES FOR PROCESS OPTIMIZATION AND CONTROL

In the view of industry leaders, unit-ops technologies are unlikely to change radically in the coming two decades. What is likely to change is how unit-ops will be managed.

Large productivity gains registered by U.S. industry in recent years are being credited in part to the implementation of IT. Our critical-technologies discussions revealed that the IT revolution is coming to the mining industry as well and will have a significant impact on the productivity of mine operations in the coming years.

While the cost of machinery used in mining has been increasing—especially for metals and coal producers facing thin or negative profit margins—the cost of information has been falling, creating opportunities to use IT to optimize the use of mining equipment and boost returns on investment. Accordingly, observed one participant, technologies to monitor and optimize mining operations are making "huge strides right now." An equipment-supplier representative asserted that electronics constituted the most important developments in the machinery his firm produced, affecting "how you run it, control it, monitor it, and utilize it."

Industry representatives identified a host of IT-based innovations that are converging to support this trend:

- Unit-ops machinery is being outfitted with a growing range of sensors, imaging technologies, and controls for monitoring and managing output.

- Emerging sensor and communications capabilities are enabling real-time and minewide remote monitoring of mine conditions and materials inventories and flows.

- The rapid decline in the cost of computing power has facilitated the growing sophistication of IT hardware and software to monitor, process, and utilize the increasing flows of mine information.

- As more components of a mine operation are brought "on-line," they can be linked together through minewide communications networks and GPS-based dispatch systems to optimize the entire mining process.

- The installation of minewide communications and data networks will enable external providers to implement a range of mine-planning and management solutions.

Information technologies in mining will have a significant impact on mine operations in the coming decades, giving mine managers and staff much greater understanding of and control over mining processes. Increases in IT capabilities also are establishing the technology base necessary to support remote and autonomous mining operations.

Unit-ops machinery such as this continuous miner are unlikely to change radically in the coming two decades. However, information and communications technologies will have a strong impact on how such equipment operates. Photo courtesy of Joy Mining Machinery.

However, the introduction and diffusion of IT in mining has been slower than in other sectors, such as the petroleum and chemicals industries, in part because the mine environment presents unique and formidable challenges:

Mining equipment moves in a three-dimensional environment; the mine environment changes as mining proceeds; the mine environment is hostile to sensitive equipment; and the individual characteristics, and hence the requirements and restrictions for IT, of different mine sites vary widely.

In the following sections, we first outline basic information technologies that are proving critical in mining operations. We then discuss some of the significant capabilities for process optimization enabled by these technologies. We conclude with a review of the present status and possible future developments in equipment remote control and automation—two of the leading objectives of IT integration in mining.

INFORMATION TECHNOLOGIES

Industry participants identified a core set of technologies that are critical to enabling IT capabilities in mining. These technologies include

- Sensors
- Position monitoring
- Ruggedized on-board computer hardware
- Advanced control algorithms
- High-capacity wired and wireless communications

Sensors

A variety of sensors are being developed and deployed for collecting data ranging from ore characteristics to equipment performance and process flows. Sensors generate raw electronic data from the physical environment of the mine that can be used for real-time process monitoring, optimization, and integration.

Several participants involved in both surface and underground mining cited the benefits of on-board sensors that are being developed and deployed to monitor equipment performance parameters such as shovel-bucket and truck-bed payloads, shovel swing angles and cycle times, vehicle speeds, and material flow rates. Advances in digital imaging combined with pattern-recognition software, positioning systems, and computing power are enabling the greater use of cameras to monitor product quantity and quality at different stages of handling.

Sensor technologies traditionally used in exploration, such as shallow seismic monitoring and ground-penetrating radar, were cited in some instances as important new tools for the development and production stages of mining. Borehole tomography, said one user, "has great application in mining" for locating deposits and improving ore-grade quality control during production. These technologies help to define more precisely rock mechanical properties, ore grade, and ore-body location on a local scale. Although the technologies are being used only in isolated cases, the participants who mentioned them felt that they would become an increasingly important part of mine process-optimization systems in the coming years. Two companies engaged in underground metals mining also indicated the importance of microseismic and rock-deformation sensors for monitoring mine environments—for example, to predict roof or wall failure.

Positioning

The introduction of the GPS has had a notable impact on surface mining, representing, according to one technology user, "one of the biggest revolutions in the last 20 years." Two levels of resolution were identified: low (approximately 10-meter) precision for tracking mobile vehicle movement of haul trucks, and high (less than one-meter) precision for precise location of individuals (e.g., during surveying) and machinery such as drill bits, shovel buckets, and bulldozer blades.

A few references were made to positioning technologies for equipment in underground environments. The constrained space within underground tunnels allows for relatively straightforward vehicle tracking via handshaking with radio or infrared beacons positioned along the travel path. Two participants also cited more-sophisticated positioning systems using gyroscopes for maintaining the proper attitude of longwall shearers and for keeping them perpendicular to the coal face, although details on this technology were not provided. The foreseen benefits mentioned included keeping the shearer in the coal seam and maintaining a desired layer of coal on the rock face.

Computer and Electronics Hardware

Increasing computing power and decreasing costs of hardware were cited as major drivers of IT innovations in mining. The cost of a computer workstation that can support three-dimensional graphics used in advanced mine modeling and planning applications fell tenfold during the 1990s, reported one executive. This computing power is now available in portable computers that can be used in the field. The decreasing cost of such equipment drove the diffusion of graphics technologies from high-value-metals producers to lower-cost operations such as kaolin and phosphate production.

Several manufacturers have developed data terminals and computers specifically for mining environments. These systems allow remote data acquisition and manipulation. Important features include improved interfaces, keyless data entry, and compatibility with various wired and wireless communications platforms.

A key challenge is that of developing or acquiring computer and electronics hardware that can withstand the hostile conditions (vibration, dust, heat, physical abuse) characteristic of the mine environment. Accordingly, our discussants noted that several essential IT hardware components—cameras, lasers, radio transmitters, and infrared beacons—are being "hardened" for use in mine environments. One participant, for example, reported the development of a cost-effective, super-hard material that will make camera lenses impervious to abrasion. An equipment manufacturer maintained that developing computers able to withstand nearly nine times the acceleration due to gravity was key to the successful deployment of dispatch systems. Another equipment manufacturer noted that breakthroughs in ruggedization have made greater application of advanced equipment diagnostics feasible.

Computer Software

Control algorithms to integrate and process data being generated by mine operations are another critical component in a minewide information network. The goal is to give machine operators on the line as well as facility managers real-time and interactive access to information needed for planning, managing, and optimizing mine operations.

The requirements of such algorithms can vary enormously, depending on the particular task. For example, powering down a piece of equipment when a sensor reaches a threshold level is relatively simple compared to automating the scooping of a shovel bucket. In general, the challenges presented in mining are quite formidable, as they involve dynamic and inconsistent physical environments. Along with unreliable communications and positioning systems, insufficient software development was among the principal causes cited for the late arrival and slow diffusion of IT in mining.

Shareware-based applications, using the Linux scripting language, for example, were seen by one technology developer as a way to increase the efficiency and speed of software development in mining.

Communications

The area of underground mine communications is undergoing rapid technological development, and consequently, several generations of communications systems are currently in use at mine sites in the United States. An important trend is the transition from section-specific to minewide communications systems.

The most basic and most prevalent underground communications system is the hard-wired mine-phone, or "squawk box," used for essential communications—typically between miners in a section and the surface. These systems require a wired connection, offer limited communications points, and are considered by some to be difficult to maintain.

A more advanced underground communications technology is the leaky feeder.[1] Leaky-feeder systems enable continuous contact between mine personnel and the surface, a capability that is particularly valuable in cases of emergency. Although a mature technology, leaky feeders are still relatively uncommon in the United States: One manufacturer estimated that they are installed in only 25 to 35 percent of underground metal mines and that installations in coal mines are far less frequent.

> Operating companies noted the contribution of minewide communications systems to enhancing productivity.
>
> • An underground operating company reported the use of leaky feeders in all its mines and noted a major benefit: the ability of personnel to immediately report breakdowns without having to walk out of the section.
>
> • A manager noted the attractiveness of a leaky feeder for his facility, saying it could reduce by up to two hours the downtime from common maintenance problems such as a flat tire.

Although best suited for voice communications, leaky-feeder systems can also transmit data in small batches. Data-based applications include monitoring the location of personnel and equipment and simple remote operations such as powering up conveyors or fans.

A more advanced-generation communications technology is high-capacity, spread-spectrum radio systems. This type of radio is in widespread use in surface mine environments for equipment dispatch, vital-signs monitoring, and remote control. It is also commonly used underground to support point-to-point (i.e., operator-to-machine) remote operations, for example of LHDs. Underground mines can also take advantage of the high capacity of spread-

---

[1]A leaky feeder uses a coaxial antenna cable with loosely braided shielding strung throughout the mine works and is connected to operators at the surface. This antenna can support radio communications between personnel and equipment up to 400 feet away and out of the line-of-sight. A leaky coax is an analogous system utilizing high-capacity, spread-spectrum radio transmission over a special coaxial antenna cable having segmented braided shielding.

spectrum radio through leaky-coax communication systems, which operate in a manner analogous to leaky feeders. Still higher-capacity and faster underground communications are possible with multichannel fiber-optics. In this variation, the coaxial antenna is replaced with a fiber-optic cable connecting strategically located radio beacon interfaces. Underground leaky-coax systems are relatively uncommon in the United States, and fiber-optic-based systems are very rare. Both are more common in Canada. According to discussion participants, the future development and extent of use of leaky coax are uncertain and likely to progress faster abroad in newer and larger operations (for example, in large-vein metals production). Standardization of system components was said to be a barrier to the technology's development and acceptance.

Ultimately, the development and diffusion of robust, high-capacity wired and wireless minewide communications networks will be necessary to support the sustained data transmission required for continuous and real-time process monitoring and optimization, remote controls, and autonomous operations.

IT-DRIVEN PROCESS MANAGEMENT AND OPTIMIZATION

The technologies described above combine to enable a minewide information-sharing capability that encompasses rapid measurement, communication, interpretation, and decision support. This capability provides the potential to monitor and control mining operations in a manner analogous to that used in fixed-infrastructure operations such as manufacturing and refining plants.

Planning

Information technologies are making an important contribution to mining by helping decisionmakers model mine development and investment choices with more accuracy over the life of the mine. This capability is proving valuable in the face of thinner operating margins, the advent of larger equipment that can reduce a mine's production flexibility, and more-restrictive land-use policies.

> Having good mine-planning tools is "absolutely critical" to success in a competitive business climate, as it enables one "to mine the right coal at the right time."
> —Coal-company executive

Recent and emerging data collection and analysis tools, such as the Geographical Information System (GIS), three-dimensional graphical representation, and computer-aided design, enable decisionmakers to quickly manipulate and understand complex spatial information that was formerly committed to paper blueprints and linen cross-sections. Such technologies give management the ability to develop and analyze scenarios based on variables such as initial mine design, operation plans, stockpile scheduling, equipment utilization, and expansion options. These capabilities speed up the planning process, allow companies to optimize capital investments in both plant and equipment, and anticipate facility closure and reclamation needs.

Many benefits of recent and emerging modeling and planning technologies were cited during the RAND critical-technologies discussions.

• Computer-aided geological and geomechanical modeling helps speed the design of initial mine layouts, support requirements, and expansions, thereby reducing up-front investment costs.

• Dispatch simulations help determine the best way to route haul roads and locate waste dumps.

• GPS-based data on material quality and location are helping mines to control the geochemistry of spoil banks and thus better plan and optimize the reclamation and closure process.

• A fire-simulation planning tool expected to become available in the near future will enable firms to minimize fire risks through better ventilation design, conduct virtual modeling of fire incidents, and design emergency-response strategies.

• Blasting simulations incorporate a number of inputs such as rock type and state; face height; hole spacing, depth, and directional deviations; chemistry of explosives; and detonation timing. These data help simulate and optimize blast parameters such as strength, fragmentation, throw distance, and collateral impacts.

• A manufacturer of roof supports is now using finite-element modeling to design comprehensive ground-control strategies in underground coal mines. Demand for this new service has been so strong that the firm opened a subsidiary company devoted solely to this business.

As mine communications and computing algorithms improve in the next few years, said one technology provider, mine-planning tools will be automated, and mine plans will be continuously updated with real-time GPS and sensor data feeds.

Dispatch

One of the major manifestations of IT integration in mining has been the introduction of dispatch systems that use GPS to monitor mobile equipment positions; to direct truck flows to shovels, crushers, stock piles, and dump points; and to optimize equipment use and material flows in real time. Another benefit of automated dispatch is the ability to better monitor and control ore grades being delivered to processing facilities.

Dispatch use appears to be approaching standard practice at larger surface operations: Nearly all relevant participants are either already using it or planning to install it in the near future. The concept is being extended to track all mobile equipment, including drills, pickup trucks, and even customer deliveries via highway trucks and rail cars.

Dispatch systems can increase equipment utilization rates by as much as 30 percent. Users have found that initial investment costs could be recovered through reduced operating costs in just weeks or months after installation. "The payoff is astronomical," especially at production-limited facilities.

—Technology supplier

Surveying

New information technologies are turning surveying activities into "a one-man show," in the words of one representative, doubling labor productivity. High-precision, dynamic GPS, for example, allows surveys to be conducted quickly and accurately from a moving survey vehicle or excavator. This reduces the need to send survey crews as well as supervisors into the mine on a regular basis and mitigates safety concerns such as trips and falls on loose rock. In addition, it eliminates the need for survey stakes, greatly improving earthmoving accuracy, particularly under poor visibility conditions, e.g., at night or during rain, snowfall, or fog.

Reflectorless laser and infrared surveying technologies enable the remote (up to hundreds of meters), rapid, precise, and three-dimensional profiling and mapping of drifts, stopes, pit walls, and waste piles in surface and underground environments. These technologies allow surveyors to work at safe distances from blast piles, highwalls, slopes, and landslides. Lasers are also being used to scan stockpiles, thus conducting inventories on a regular basis. Three executives said that GPS- and laser-based surveying technologies were helping their operations better establish the proper contours and gradients needed for reclamation.

Electronic survey data can be entered into mine information systems, such as GIS databases or planning tools, and automatically updated during earthmoving operations.

Laser-based survey technologies can be used to map a drift at 3,000 times higher resolution and 100 times greater speed than manual methods.

—Technology supplier

Equipment Positioning

Integration of high-precision (one-centimeter accuracy) GPS-based surveying, equipment positioning, and ore-body maps is enabling managers to automatically monitor and direct earthmoving activities such as shovel-bucket and bucket-wheel scooping; blast-hole drilling positioning, alignment, and depth; ore stacking; and bulldozer grading.[2] One of the principal benefits of this technology is a greatly reduced dependence on manual surveying. Once an initial survey is conducted, information systems can update maps in real time as earth is moved, and updated information can be continuously shared throughout the mine site. For example, monitors in operator cabins can be fed continuously updated, color-coded maps showing topography, locations of ore and waste, and tool positions.

In addition to decreasing surveying requirements, increased precision can result in better blast control, decreased ore dilution, more uniform bench heights, and greater operating speed. This can generate large productivity increases: "Customers typically see significant improvements on the order of 25 percent," said one supplier. A manufacturing firm claimed that its customers realize 30 percent improvements in productivity from these technologies. According to literature provided by a technology supplier, one mine estimated a 90 percent reduction in survey time from implementing GPS-guided bulldozer grading.

High-precision GPS positioning appears to be growing in popularity, with initial installations preferentially applied to drilling systems. Overall, they are still in relatively limited use, however; one manufacturer estimated that high-precision shovel positioning was in use at about 30 sites worldwide.

Performance and Productivity Monitoring

Performance monitoring integrates data from equipment sensors with variations in operating-costs to quantify the effects of different operational choices.

Parameters such as number of vehicles, haul distances, maintenance scheduling, operating hours, loads, swing angles, cycle times, particle sizes, speeds, flow rates, and operator scheduling can be compared with various productivity metrics such as labor, maintenance, consumable costs, downtime, and ore-production rates to help make informed decisions about how best to operate a particular mine or production technology. Productivity monitoring is particu-

---

[2]GPS is available at a range of resolutions. Standard autonomous receivers have an accuracy of one- to 10-meter resolution. Differential GPS, which uses base reference stations or pseudolites (to correct for variations in the GPS signal), has accuracies down to 20-centimeter resolution and is used for dynamic positioning and automated dispatch. Real-time kinetic provides one-centimeter-accuracy GPS positions and is used for surveying and machine guidance.

larly beneficial for situations where mine operations vary with time due to expansion, addition of new technology, or transient variations in product demand. One equipment supplier noted that his company uses productivity monitoring to analyze repair claims. Another supplier reported using performance monitoring to verify the legitimacy of warranty claims.

Operating-company and technology-supplier representatives illustrated many uses and benefits of performance monitoring.

- Through the use of bar-code readers to input operator ID, a coal-mining operation monitors equipment performance by work shift. Management can ascertain when equipment is turned on and off, equipment performance, and volume of coal produced per shift.

- Unsatisfied with the performance of its new investment, a metals producer scrapped its vertical impactors (a newer technology) and reverted to the use of cone crushers (an older technology) outfitted with enhanced control systems that automatically optimize hydrostatic pressure and close size and thus maximize crusher throughput and availability. The three-stage crusher circuit is run from a computer screen by one operator.

- A metals mining firm has developed an implemented system of "Critical Performance Indicators," such as unit costs and availability, labor productivity, and safety. These indicators (which are derived from a series of measures deemed critical by management) are then compared with company goals over time and across units.

- A materials producer is developing a companywide GIS database that incorporates data on each of its mining units, including production, costs, orders, and inventory. By putting real-time process data on every manager's desk, the company hopes to better manage inventory, improve load management between facilities, track unit progress in meeting goals, and facilitate benchmarking.

Finally, several discussants pointed to the goal of integrating the local data networks of mines and quarries within an organization and with outside vendors and partners. The vision, explained one discussant, is "to effectively make any piece of mining equipment a node on the internet." Once this occurs, an unlimited range of outside providers can tap into an enterprise's data network and "layer in" process-optimization solutions for mine planning and management, blast optimization, equipment performance monitoring, equipment dispatch, inventory management, regulatory compliance, training, etc.

Process Integration

The availability of minewide information sharing has provided the capability to begin linking previously separate operations around the mine such as surveying, mining, processing, and reclamation (see Figure 4.1).

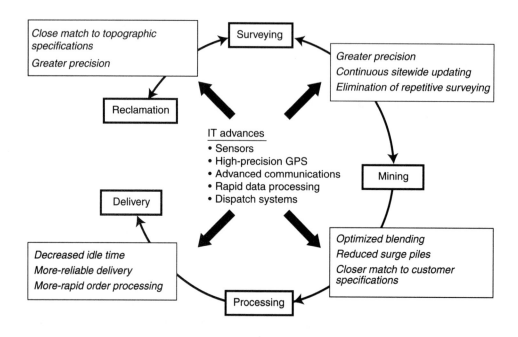

Figure 4.1—IT Applications for Mine Process Optimization and Integration

As discussed above, surveying can now be integrated with mining through continuous map updating as mining proceeds. After an initial survey, high-precision equipment positioning and material tracking allow topographic and ore-body maps to be automatically updated and distributed throughout the mine with each shovel scoop. This integration can extend through the life of the mine and into the reclamation stage, where high-resolution earthmoving systems can be used to optimize restored topography.

A second aspect is the closer integration of ore production and processing operations. One of the most important technology innovations in mining, one executive said, was the use of GPS to identify material as it is mined and the use of in-process sensors to track the material's movement in the handing process. Similarly, geomechanical and geochemical information collected immediately in front of the advancing face in underground operations is being used to plan specialized blasting, roof-support, and ore-processing needs.

With such capabilities, mining sequences can be planned and modulated to add flexibility to and optimize plant operations. For example, information from the mine, such as ore production rate, ore grade, and particle size and hardness, can be monitored and used to adjust ore delivery; crusher power; blending ratios; stacking, leaching, and washing rates; and other processing variables. Conversely, stockpile size, specialized blending requirements, and changing customer specifications can be communicated upstream to the mine face so that operations can be adjusted accordingly. Such capabilities are especially important for operations such as quarries and industrial minerals producers that want to ramp up production to meet strong demand.

Surface imaging and detection technologies are gaining wider application in mining.

• A metals producer is deploying video cameras and image-recognition systems to monitor ore going into primary crushers, at the first transfer points, and at the mill.

• To reduce the potential for contamination of industrial minerals, a producer installed video cameras inside its product storage bunkers to remotely monitor inventory levels and operating equipment.

• Automated measurement of material drawdown at crushers is enabling an aggregates producer to optimize ore flow rates and maximize crusher throughput in a heavy-demand environment.

• A coal company has refined and employed high-resolution surface seismic monitoring to study geologic anomalies between core holes in advance of mining that may affect coal recovery or ground control. When areas of concern are detected, additional core holes are drilled to study the suspected anomaly in greater detail.

The use of advanced information systems to optimize ore grades becomes more important as the number of shovels and excavators in a mine falls, reducing equipment operators' ability to blend ore grades, as one observer noted. Likewise, as downstream processing and utilization become more sophisticated and rigidly controlled (in response to environmental regulations, for example), optimizing and maintaining quality control of feed stocks become more critical.

Knowledge Management

Many discussants noted that although mine operations are generating more data, such information rarely is well utilized. For example, two equipment and service suppliers said that while their firms' technologies generated large volumes of data, the information is never reviewed except in summary form or in the case of an exceptional event. Echoing the view of several discussants, a manager argued that the challenge is, first, to decide what information is important, and then to decide how to make use of it.

The newest haul trucks provide a large volume of performance data. The information (presented as codes and numbers) may be used by specialists well versed in data analysis, but it is not readily interpreted by on-site mechanics. While these trucks record the number of times an engine exceeds a specified RPM, the time of each incident is not recorded, and thus the operator cannot be identified.

—Mine manager

Accordingly, another critical technology is effective knowledge management: tools and capabilities for distilling complex mine information into an actionable format that a mine engineer or operator in the field can comprehend and act upon in real time. One means to improve data utilization, said an operating company representative, is wireless data-transmission systems to get the information off the equipment and to the proper location to analyze it. On the other hand, it was argued, such data typically require complex and time-consuming interpretation that can overwhelm mine operators. One solution cited is the use of simple graphical interfaces, for example, red and green icons indicating where an operator should go.

"Data is not as interesting as insight."

—Coal-company executive

Several discussants argued that operating companies were slow to realize the benefits of knowledge management. "What will all this data do for us?" a technology supplier asked rhetorically. Many technology providers argued that mining companies do not have a good understanding of cost centers across their entire operation. This can deter investment decisions, they suggested. One equipment provider argued that it was difficult to convince operating companies that his firm's technologies could radically reduce their ore-handling costs.

"There is more to be gained with information technology over the short term than automation."

—Technology supplier

Knowledge management appears to be correlated with company size: The larger a mining organization is, the more resources and know-how it can dedicate to gathering and interpreting operational data and discerning how to utilize the information in the field. With their greater analytic capabilities, one technology provider observed, larger mining companies tend to be more cost conscious and to more closely evaluate technology investments and operations and maintenance (O&M) costs.

REMOTE CONTROL AND AUTOMATION

Remote control (also referred to as telemining) and automation have been high on the mining R&D agenda for a long time. During our critical-technology discussions, proponents of these capabilities cited numerous anticipated benefits, including

- Higher equipment utilization by avoiding lengthy shift changes and transits (up to three hours in some cases), breaks, and worker fatigue.

- Reduced need for human support systems (most importantly in underground environments).

- Reduced wear and tear on equipment.

- Increased safety resulting from moving the operator away from an active mine face.

Although they are based on the same set of core technologies, remote and autonomous operation are different in some fundamental ways from the process-optimization capabilities discussed above. The primary distinction is that with remote control and automation, computer algorithms are used to control actual equipment *operations,* rather than operational *decisions.* Participants noted that this new paradigm amplifies existing technological challenges such as development of robust algorithms and reliable communications, and also introduces a suite of new hurdles, such as operating with different human sensory inputs, increased reliability requirements resulting from the absence of human assistance, and difficulties of integrating remote or autonomous and manual equipment at the same facility or section.

> "You can't automate something that doesn't work properly and reliably."
>
> —*Coal-company executive*

> "Before you automate something, you must make it efficient and you must understand the system. . . . The technologies for driving machines have been around for a long time. But to do this with high reliability and integrity 24/7 is very hard."
>
> —*Technology supplier*

Another impediment is cost: The unique character of each mine operation requires that remote and autonomous systems be highly customized. "We never build the same thing twice," observed a coal-company executive. These systems also require advanced technological capabilities, particularly remote ore-characterization and horizon-monitoring sensors, that are not yet commercially available and that few suppliers have the capability to develop and integrate. The systems will require that mines develop and promote talent in disciplines outside of traditional hiring and promotion paths (e.g., computer programming, systems operation, and electronics). Finally, developing fail-safe systems (for example, assuring that radio communications do not interfere with other operations) and ensuring the safety of personnel working in an environment of autonomous machinery were identified by some discussants as remaining technology development challenges. Such challenges ensure that the advent of the "workerless mine" will be beyond a 20-year horizon.

> "We've had a robotic mail cart in this building for years. Robotic technology is available; it's just not advanced enough to cope with the complex mine environment."
> —*Technology supplier*

There was no strong consensus among the discussion participants on the feasibility or benefits of remote and autonomous equipment operation. We heard from many ardent proponents: "The payoff is astronomical," claimed one advocate. The representative of an operating company with considerable experience in prototype development claimed that the commercial application of remotely controlled equipment was likely to double the productivity of the company's underground mines—an improvement that will outpace any other technological innovation in the company's 100-year history (see Figure 4.2). He concluded, "Remote operation can never beat the best human operators, but it can always beat the average." On the other hand, several participants noted that many of the anticipated benefits of remote and autonomous operations had yet to be convincingly demonstrated; as one caustically observed, "Imagine a doctor operating by remote control."[3]

The lack of consensus may stem from differing views of the role of mine personnel in a remote or autonomous mining operation. Proponents claim that the technologies will reduce the need for human support systems, such as ventilation, in underground facilities. Yet several speakers observed that personnel will be required to remain nearby to service the equipment. Another discussant noted that laser guidance systems and image-recognition systems will require

---

[3]As a matter of fact, remote-controlled and automated surgical procedures are already being envisioned and performed.

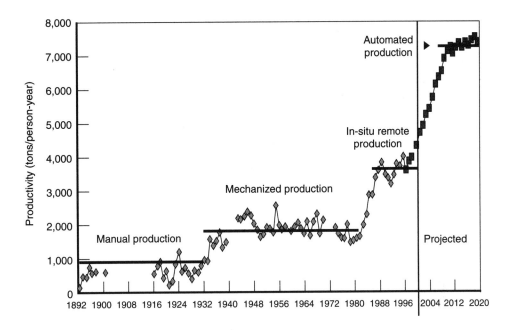

Figure 4.2—Past Productivity and Anticipated Productivity from Technology Change in One Company

ventilation to reduce dust and smoke. Autonomous haul trucks are likely to have personnel on board, at least initially, overseeing operations, according to another discussant.

Remote Control

Use of remote control is standard in only a few applications in the United States, according to the technology-discussion participants.

Line-of-sight wireless remote control is routinely used for guiding LHDs in unsupported stopes in underground metal mines. Several participants noted that remote control enabled the development of large, unsupported stopes and full extraction in mucking out. With remote control, some personnel are able to operate an LHD beyond their line-of-sight, using sound cues. The addition of remote video monitoring provides enhanced "around-the-corner" mucking ability, requiring fewer draw points to be cut per stope. Many underground coal continuous miners utilize untethered remote controls, where the operator stands behind the machine as it cuts. Highwall coal mines also drive entries using remote controls.

> Remote-controlled LHDs have allowed a gold producer to recover more ore per stope, even as overall production volumes have fallen. The cost of conversion for each unit ($30,000) was equivalent to the cost of two cone muck piles. When there has been a stope fall, the equipment was dragged from the muck pile, repaired, and returned to work.
>
> —*Gold-company executive*

Several participants expressed a strong interest in extending remote-control capabilities to cover more mining operations, despite their relatively limited use to date. Three participants pointed to advanced remote-control capabilities being developed on an operational basis in Sweden, where an individual can remotely operate three LHDs, controlling the scooping function while an automated navigational system takes over and controls the tramming and dumping functions.

A mining-company/manufacturer/government partnership in Canada has made considerable progress in the development of a fully remote underground mining system, the progress of which, several study participants noted, they are following closely. The consortium's goal is to run all development and production operations from the surface at greater efficiency and less cost than could be done with traditional in-situ methods. Thus far, they have developed production drills that are remotely operated in working mines. Long-distance wireless remote control of other operations, such as surveying, development (jumbo) drilling, blast detonation, LHD mucking and tramming, and shotcrete application, has been demonstrated with varying degrees of success but is not yet in commercial use.

One finding of this effort is that the need to provide operators with multiple sensory inputs may not be as important as expected: Remote operators are provided with sound, but they routinely turn down the volume until a visual cue prompts them to listen, according to one observer.

Automation

No fully autonomous mobile equipment was reported to be in use in the United States. The only fully automated task reported to be in relatively widespread use was automatic advance of shields and face conveyors in longwall coal mines.

Automation should be viewed as an evolutionary process, said one executive. Thus, considerable attention has been focused on developing semiautonomous and operator-assisted controls as an intermediate technology. Operator assists (or "smart" technologies) have been developed for a range of applications, and there was a strong consensus among the participants that their development and diffusion is an important, albeit less noticed, trend. Examples of demonstrated or commercially available operator assists cited by study participants include surface and underground drill-bit positioning; control of drilling speed, pressure, and depth; shovel, wheel loader, and LHD scooping; and continuous-miner guidance. Autopositioning of loaded shovel buckets over truck beds is expected to be available in 2001. According to a truck manufacturer, other tasks that are under consideration for operator assist include truck steering, dumping, collision avoidance, and ramp-climbing.

Operator's view of a 50-cubic-yard shovel bucket scooping overburden. Heavy excavators are equipped with operator assists to optimize scooping power and reduce wear and tear on the machinery.

Several types of autonomous mining equipment will be available on a commercial basis within the next few years, said several equipment manufacturers (see Table 4.1). Many participants expressed interest in the advent of autonomous haul trucks. Although hauling was generally cited as being one of the most difficult tasks to automate, the large number of trucks in use makes the potential payoff to both mining companies and manufacturers significant. One mining executive noted that labor accounted for 30 percent of haulage costs at his operation and that the company had looked into acquiring driverless technology; however, the control systems were deemed to be inadequate to reliably implement a driverless system at that time.

Table 4.1

Anticipated Availability of Autonomous Mining Equipment

Equipment	Anticipated Date
Autopositioning shovel bucket	2001
Autonomous surface-haul truck	2002–2005
Autonomous LHD vehicle	2005
Autonomous surface drill	2003
Autonomous shovel	2005

Source: RAND discussion participants.

The discussants differed widely in their opinions on where automation was likely to proceed most rapidly; this difference is most likely a function of the diverse nature of mining. Automated drilling and surveying are proceeding more quickly in open cast mines, due to the availability of GPS and wireless communications in the open environment. Semiautomated tramming of haul trucks has been introduced in several underground mines—a development facilitated by the more confined space, uniform lighting, and stable operating environment. In an open pit, by contrast, more precautionary measures must be integrated into vehicle controls, to prevent haul trucks from deviating from their course or driving off a bench. On the other hand, one industry representative contended that the application of driverless trucks might proceed more rapidly in quarries, given their shallower profile and the less-dynamic nature of the working environment.

Chapter Five

CRITICAL TECHNOLOGIES FOR OPERATIONS AND MAINTENANCE

Given the large expenditures on capital equipment characteristic of mining, technology developers and users alike place a very high priority on improving equipment productivity. Proper maintenance is essential to maximizing both availability and productivity. At the same time, maintenance services are a major cost element of all mine operations.[1] Accordingly, after process optimization, equipment operations, maintenance, and repair technologies were most commonly cited as critical by the participants. The prominence of O&M technologies in mining coincides with earlier RAND assessments of critical technologies across U.S. industry.[2]

Maximizing returns from fixed capital has become increasingly important to the success of coal and metals producers as margins have been squeezed by competition and weak commodity prices. Moreover, as individual unit-ops increase in capacity (and, for example, the number of shovels and trucks in a mine is reduced), outages have correspondingly greater impacts on overall mine production. Similarly, as mine processes are more tightly integrated (as in a longwall), an outage in a single mission-critical component can shut down an entire production line. Optimizing O&M is likely to become even more critical as mining equipment and geological conditions become more complex and the provision of support for skilled maintenance staff in remote locations becomes more difficult.

Maintenance is especially critical for the stone and aggregates industry, as producers strive to keep up with strong demand—a trend that is likely to last well into the present decade. When asked to identify bottlenecks to boosting pro-

---

[1]To illustrate the scale of the issue, one participant estimated that maintenance alone accounted for 25 to 35 percent of a mine's operating costs.

[2]See Steven W. Popper et al., op. cit.

ductivity at his firm, a stone and aggregates executive ranked insufficient maintenance as his second most important organizational challenge.

A loaded 240-ton haul truck climbs out of an open-pit mine. Reliability becomes more critical as equipment increases in scale, because an outage affecting a single piece of machinery can have a large impact on the output of a mine or quarry.

To avoid such bottlenecks, several priority O&M solutions currently are being developed and applied in mining:

- Equipment monitoring and diagnostics

- Repair and maintenance scheduling

- Maintenance technologies and practices

- Engineering robust systems

In addition, mining and quarrying enterprises are increasingly outsourcing their maintenance operations (this discussion is reserved for Chapter Six). While most of the participants in the RAND discussions stressed the importance and criticality of O&M, the practices they engage in and their views about solutions varied tremendously.

EQUIPMENT MONITORING AND DIAGNOSTICS

Many technology providers highlighted a variety of recent and emerging technologies for equipment performance monitoring and diagnostics.

Major pieces of equipment are being outfitted with an increasing range of on-board sensors to monitor the state of critical systems—a capability one executive described as "absolutely critical." Vital-signs monitoring outputs include fluid temperatures, levels, and pressures; engine speed and gear position; brake temperature; bearing rotations and temperature; drivetrain performance; and hours of operation. By 2001, new-generation heavy-truck tires will have sensors that monitor tire pressure and temperature and transmit the data to on-board or roadside receivers. Real-time monitoring is expected to help operators reduce tire wear, improve equipment performance, and avoid catastrophic tire failures. Additional sensors, such as on-board oil analysis and structural-strain gauges, are under consideration or in development.

> Remote monitoring of amperage levels on conveyor belts at an underground coal operation has reduced the number of staff required to watch the belts and provides central dispatch with real-time notification when a belt goes down.
>
> —*Coal-company executive*

Vital-signs data typically are manually downloaded at regular intervals for analysis and are entered into a spreadsheet or analyzed as a shift report. In some cases, monitoring data are transmitted in real time via radio frequency or wire to a remote location, such as the mine's dispatch center. For example, programmable logic controls provide mine operators with performance and diagnostic data on belts, fans, drive motors, power distribution systems, etc.

Off-board diagnostic techniques include vibration, lubricants, ultrasonic, and thermographic analysis. Many mining operations reported regularly analyzing their engine oil, hydraulic fluids, and other lubricants for signs of wear, contamination, or malfunction. High soot and fuel content, for example, is an indicator of piston blow-by and lower engine output. Fuel or coolant contamination, in turn, indicates lubricant degradation and reduced lubrication capacity. Progressive analysis over time can indicate equipment abuse or the need to modify the lubrication or maintenance regime. Vibration analysis of critical elements (e.g., bearings, bearing races) is assisting the implementation of predictive maintenance regimes, and mission-critical systems increasingly have real-time vibration sensors built in.

```
                          SHIFT REPORT

Site:                          Driver: Driver
Vehicle No.: 268                 Type:
Data File:   08120807.dat      Version:
From:        08/02/98  10:30AM      To: 08/02/98  04:09PM

                       PRODUCTION REPORT

Total Tonnes                  247.0 tonnes
Total Bucket Loads             50 loads
Average Bucket Load            4.9 tonnes
Average Bucket Load Factor     49.4 percent (10.0 tonnes = 100%)
Average Over Load Weight       10.6 tonnes  ( > 100% )

                          TRENDING
Total Card Hours               5.6 hours
Total Engine Hours             5.6 hours
Engine Utilization             99.9 %

Travel Distance                1.8 km
Travel Time                    0.7 hours
Avg. Speed                     2.4 km/h
Avg. Cycle Time                5.4 minutes

Hydraulic Oil Temp             67 C
Transmission Oil Temp          51 C

Fuel consumed                 125.2 liters
Average fuel consumption/hour  22.2 liters/hour

Avg. RPM                      1653 RPM
Avg. Engine Oil Pressure       370 kpa
Avg. Engine Oil Temperature     84 C
Avg. Engine Air Temperature     52 C
Avg. Engine Output Torque      452 Nm

                       HIGH GEAR USAGE
                -------Number of Shifts-------
               Loaded       Unloaded        Total
1st  Gear        48           141            189
2nd  Gear         9            42             51
3rd  Gear        15            60             75
4th  Gear         0             0              0

                     ALARMS/EXCEPTIONS

Event Description              Duration  No. Events   Max/Min Value
                               (min.)
Bucket Overload > 10.0 tonnes     1        10          13.0 tonnes
Hoist Rod Pressure > 2070 kpa     9         6          2091 kpa
Hoist Base Pressure > 20685 kpa   0         0         20000 kpa
Hydraulic Oil Temp > 116 C        0         0            89 C
Transmission Oil Temp > 110 C     2        15           126 C
Engine RPM > 2300                 0         0          2200 RPM
Engine Oil Pressure < 138 kpa     0         0           144 kpa
```

A shift report downloaded from an LHD. Such data reports can help mine management monitor the vehicle's output and efficiency, work-shift performance, and malfunctions.

Vibration analysis and lubricants analysis were cited as two important O&M technologies.

• A mine manager reported that monitoring crankcase oil for brass—an indicator of bearing wear—helped his unit avert costly engine rebuilds.

• A major aggregates producer reported that vibration analysis enabled technicians to accurately predict when equipment would go down and was a more reliable gauge of performance than the accumulated knowledge of plant foremen garnered over time.

• The maintenance service group of a heavy-equipment manufacturer reported using ultrasound and vibration analysis to predict consistently and with good accuracy structural-steel failures as far as six to seven months in advance.

• A shovel manufacturer analyzes the vibration of gear assemblies at the factory and supplies its customers with a "birth certificate" that provides users with baseline reference data to monitor trends in the condition of the assembly over time.

REPAIR AND MAINTENANCE SCHEDULING

Maintenance practices vary greatly throughout the industry and do not appear to be correlated with firm size or industry segment. Many mining companies follow manufacturer specifications, while others "run their equipment until it drops," said an industry observer. "Every part we change has gone to destruction," reported an operating-company manager. Many executives noted a shift to predictive maintenance (also referred to as reliability-centered or performance-based maintenance), whereby servicing is conducted according to machinery and equipment performance parameters and the needs of the mine operation, rather than manufacturer guidelines or rules-of-thumb.

Predictive maintenance is supported by enhanced vital-signs monitoring that signals when maintenance or repair interventions should be taken. Such advanced diagnostics can flag a problem much earlier than conventional indicators, such as noise, smells, or power loss can. (As one industry executive observed, a noisy bearing indicates that too much damage already has occurred.) Alternatively, variations in performance over time are compared with historical data or design specifications to help schedule maintenance most efficiently.[3]

The keys to success with such diagnostics, as one observer noted, include obtaining sufficient trend data over time and developing the confidence of equipment operators and maintenance personnel in the predictive abilities of what at first glance looks like abstract data. According to its advocates, predic-

---

[3]Historical data ("working curves") may be developed either from the mine's own maintenance records or from a variety of sites observed by contract maintenance services.

tive maintenance results in less-frequent and less-serious equipment failures, less unscheduled downtime, and enhanced safety, and thus can be viewed as a risk-management tool. However, another executive added that he did not have enough data on hand to determine whether his company's predictive maintenance regime was contributing to higher availability. Two other executives noted that the challenge of cost-effective maintenance lies in determining which parts can be run to destruction and which parts are critical and require predictive maintenance.

> Predictive, or performance-based, maintenance is being practiced widely in the mining and quarrying industry.
>
> • One large mining firm described its current preventive-maintenance implementation efforts as "scatter-shot," with the goal being more targeted monitoring and greater fact-based analysis. The goal of the maintenance-improvement process was to move equipment availability rates from the "high 80s to the mid-90s."
>
> • A large underground mining firm uses performance data to schedule maintenance during planned downtimes, such as at the end of a panel in a longwall operation, when equipment must be dismantled and moved. By carefully scheduling maintenance, operators can better "define the terms of engagement," thereby creating a better work environment and increasing safety.
>
> • Wheel-bearing temperature indicators being installed on large haul trucks warn of imminent problems, enabling the driver to decide to schedule maintenance or, in serious cases, to stop the vehicle immediately. Such systems can reduce major bearing damage, and hence maintenance costs, by 90 percent.

The maintenance benefits of on-board diagnostics extend beyond scheduling. Equipment health data can be transmitted to a mine-site control center, a regional headquarters, and the manufacturer. This allows maintenance actions to be planned, parts ordered, personnel notified, and production operations modified in accordance with performance parameters.

MAINTENANCE TECHNOLOGIES AND PRACTICES

> New-generation 360-ton haul trucks incorporate modular designs to facilitate the change-out of hoses, components, and engines. According to one operating company, this represents "a big breakthrough in serviceability."
>
> *—Technology supplier*

Equipment suppliers are making improvements in equipment design to facilitate maintenance—for example, centralizing the locations of maintenance ports and service elements and making them easily accessible. Key systems, such as planetary gear assemblies and engines, are being bundled as modular "plug-and-play" components that can be easily accessed and quickly replaced. Modular design and subsystem replacement can reduce in-field maintenance time, reduce the need for highly skilled workers and clean environments in the field, and allow more time-consuming and complex repairs to be accomplished under optimal conditions.

Many mining operations are investing in better maintenance areas, greater contamination control, more thorough and efficient record-keeping, and more complete and careful equipment rebuilds. As mining equipment gets larger, components requiring service, such as wheels, drivetrains, and bearings, also increase in size, motivating companies to develop new maintenance capabilities.

> A repair technician devised a novel method that uses suction to speed crankcase oil drainage, thereby reducing the time required to a fraction of the standard three to four hours.
>
> *—Gold-company manager*

Enhanced diagnostics offer the ability to quantify the effect of deferred maintenance and nonstandard equipment operation. Several participants noted that mines often operate equipment outside of the manufacturer's recommended conditions (e.g., they overload trucks, ignore warning lights, etc.) in order to meet production targets. Discussants noted that it is often difficult to determine the financial costs of such behaviors. The integration of O&M data is allowing managers to quantify the costs of stressing equipment, although the limited availability of high-capacity underground wireless communications systems has generally limited these capabilities to surface environments.

ROBUST SYSTEMS AND MATERIALS

As noted in Chapter Three, equipment at aggregates operations is being placed under particularly demanding conditions to meet market demand. The development of more-robust subsystems is helping to reduce costs by lengthening maintenance cycles and reducing the incidence of critical failures.

> Using more-expensive, longer-life parts, such as hardened excavator bucket teeth, entails a significant initial cost premium but returns greater savings over the long term.
>
> *—Operating-company manager*

One operating-company representative noted that a principal determinant of equipment life is the capacity of lubrication systems. The heat tolerance of lubricants is being increased, reducing cooling requirements of engines and motors. The resulting ability to use lower-viscosity (thinner) lubricants in drive trains increases the transmission of useful energy to the road and rock face, reducing engine wear and tear. At the same time, the soot-holding capacities of motor oils have increased, lengthening the intervals between oil drains by 100 percent and reducing air emissions, said one manager. According to another observer, large oil reservoirs and "superfiltration technologies" such as centrifugal dirt separators can extend maintenance intervals. Oil changes on haul trucks are typically needed after every 500 hours of operation, with the latest super-size trucks able to run for 1,200 hours. The interval should increase to 5,000 hours in the future, this observer said.

Key innovations are the rise of maintenance-free systems such as "filled-for-life units," currently available in some noncombustion applications, e.g., bearings and gear boxes. Another innovation is automatic replenishing systems, such as oil burners that inject a fraction of crankcase oil into the engine combustion chamber.

The development of robust component systems was cited by one industry executive as an important prerequisite for the successful implementation of remote-controlled and automated equipment, given the expected distant location of support personnel and time lags in responding to equipment malfunctions.

Chapter Six

CRITICAL TECHNOLOGIES FOR ORGANIZATION AND MANAGEMENT

A common thread throughout our discussions was the importance of good operations management to success in a very competitive market environment. Mining is not unique: Since the 1980s, efforts to improve management (epitomized by benchmark companies such as General Electric and Intel) have been undertaken throughout the business community. Improving mine organization and management practices has the potential to greatly impact productivity, staffing patterns, health and safety, and even the layout of mine and quarry operations.

Unlike hard-technology solutions, the critical management and practice innovations cited by study participants tended to be unique to each company or operation. Areas of technology-related management innovation range from a focus on the individual to reorganizing the operations of an entire site and include

- Outsourcing of important mine tasks

- Health and safety

- The human environment

- Technology and human-resources management

OUTSOURCING

Our discussions revealed that while fundamental changes in unit-ops technologies in the coming years are likely to be rare, the way in which mining equipment and services are being acquired, operated, and maintained is changing dramatically. Mine activities that commonly are outsourced include drilling and blasting operations, equipment-performance and regulatory-compliance monitoring, warehousing, and maintenance and repair. On the other hand, the outsourcing of an entire mine operation—commonly referred to as contract mining—while frequent in the industry, was not looked upon favorably by our discussion participants.

Outsourcing of maintenance and repair was noted by many participants in the technology discussions as an industrywide trend. Maintenance agreements vary considerably and are influenced by the size of the mine unit and the availability of off-site maintenance facilities and personnel. But increasingly, mining companies are shipping components or entire pieces of equipment to suppliers or maintenance and repair specialists off-site instead of supporting a comprehensive maintenance operation and performing major repairs on-site. This outsourcing can be further enhanced by transmitting equipment diagnostic data directly to manufacturers, service contractors, and parts suppliers, allowing maintenance actions to be scheduled, parts ordered, personnel notified, and production operations modified automatically. The development of such a remote-service infrastructure has been facilitated by improved logistics and transportation to mines (particularly in the western United States) and is beginning to be used by all segments of the industry. More comprehensive O&M contracts entail providing dedicated personnel on-site along with the necessary equipment and are typically used by large operations with heavy maintenance burdens.

Benefits of outsourcing that participants cited include

- Lower O&M labor costs
- Lower maintenance-facility and overhead costs
- More-predictable maintenance expenditures
- The ability to obtain the necessary capabilities at mines where they are not locally available

An additional advantage of outsourcing, according to a technology provider, is that mining companies often can draw on the know-how accumulated by specialists from operations both within and outside mining (e.g., engine or lubricant manufacturers). Finally, by moving maintenance functions off-site, mining companies can ease their regulatory-compliance burdens.

Some participants mentioned disadvantages and impediments to O&M outsourcing: small mine-unit size, managing the logistics of several maintenance contractors operating on-site, lack of convenient service personnel and facilities off-site, mistrust of contractors or the feeling that they are not members of the "mine team," and questions of union acceptance. An exception to the trend toward contract maintenance is found in the underground coal-mining sector. One participant with broad industry experience noted that the coal producers prefer to cross-train their operators to run equipment and repair it themselves.

A motivation for this is flexibility: In a mine with a single longwall panel, for example, a breakdown must be repaired immediately.

As outsourcing develops, some mining companies are focusing on core competencies, while others are developing new lines of business.

• A coal-company executive cited warehousing as one of his firm's top three cost centers and reported efforts to "push everything back to vendors" to reduce on-site warehousing requirements and inventory overhead through just-in-time parts delivery.

• In response to the greater use of mobile crushers across the aggregates industry, one aggregates firm has made efforts to develop a major capability in overhauling mobile crushers so that it can provide this service to other firms as a separate line of business.

THE HUMAN ENVIRONMENT

Assuring the health and safety of mine workers was an important concern of the study participants, and many noted that health and safety figure prominently in their statements of company objectives. Many mining executives claimed that their operations exceeded regulatory requirements for health and safety and also exceeded average industry performance.

When asked to identify critical health and safety technologies, many operating-company representatives focused on personal protective equipment and innovations they or their organizations had pursued.

• All personnel at one firm's underground coal mines are required to wear reflective safety vests—a measure described as "a simple but effective" way to reduce injuries.

• In collaboration with a glove manufacturer, a coal company developed and deployed Kevlar-reinforced, padded metacarpal work gloves after observing a high incidence of injuries due to pinching and abrasion. The gloves "significantly reduced" the severity of hand injuries.

• A communications company has deployed a warning system in which the miner's radio is integrated with the helmet-lamp battery. In the event of an emergency, a warning signal triggers the lamp to begin blinking. The miner can then respond, providing instant confirmation of his or her location and status.

• A large metals producer developed its own catalog of recommended personal safety equipment, with an emphasis on low-tech, wearable solutions, including gloves, respirators, and air hoods.

Drivers of enhanced health and safety efforts cited were risk management (e.g., avoiding lawsuits and litigation), reducing lost worker time, improving productivity, maintaining morale, and common sense. Complying with health and safety regulations appeared to be a secondary technology concern. For example, when asked to identify critical innovations for mining health and safety, discussants rarely mentioned compliance technologies.

Rather, they typically pointed to what they saw as innovative personal safety equipment or programs their firms had undertaken on a voluntary basis that went beyond regulatory requirements. For instance, a coal-company representative reported that his organization replaced the ladders on its loaders with steps, which were seen as safer and more user-friendly. Another coal company expressed its interest in obtaining a new earmuff technology that monitored and recorded ambient and transmitted noise levels, allowing a quantitative assessment of effectiveness.

Several of the study participants noted that injuries in mining are increasingly resulting from human errors rather than equipment failures or geologic factors. Consequently, in conjunction with improving equipment and environmental safety, the mining community is emphasizing the importance of safety through behavior modification. In an effort to reduce strains and sprains, a coal company encourages its workers to stretch for 15 minutes before a shift. Two firms reported the implementation of companywide accident-reporting systems to benchmark performance across all of their units (regardless of the type of mining) and identify problem areas. Another firm implemented an accident- and violation-reporting program in which memos discussing accidents as well as rule infractions not resulting in injury are circulated to all company personnel.

In many cases, efforts to improve health and safety entail behavioral and practice innovations rather than technical solutions.

• An international company has developed a companywide safety index composed of four criteria measures, two of which are changed every month to keep the effort fresh.

• Under the motto of "creating safe workplaces and caring for the individual," the management of a major surface mine periodically observes the work of rank-and-file staff at the job site and presents its findings through personal, informal communications. The observations (made by foremen as well as the company president) increase worker safety awareness, empower staff through the exchange of ideas and experiences, and help "break down the traditional labor-management hierarchy."

• Management at a large surface mine has its haul trucks drive on the left side of the road to move the operators further from danger in the event of a collision.

• As part of an effort to improve ergonomic conditions in its underground operations, a major coal producer in the early 1990s instituted a program to observe and modify work practices and conditions. The program reportedly resulted in a reduced incidence of strains, sprains, and back injuries.

Underground mine safety is benefiting from improved geomechanical monitoring, said several industry leaders. Microseismic and rock-deformation sensors are used to monitor pillar deformation, roof sagging, and crack propagation; to identify unstable conditions; and to help predict dangerous occurrences of catastrophic brittle-rock failure (bumps or bounces). Several participants noted that these concerns are expected to increase as U.S. mines begin to operate at greater depths in the future.

The increasing reliance on O&M outsourcing, noted by many discussants, also can have important health and safety implications. For example, an operating company may elect to outsource blasting to reduce the risk to its employees. While this may appear to represent a displacement (or offloading) of risk, two technology providers argued that the outsourcing of blasting to specialist firms with specialized training and equipment and an organizational focus on blast safety was in fact a move to an overall higher level of safety. "Our people are trained better than anyone else in the world," said one blast-services executive.[1]

A video camera mounted on the front of a haul truck at a mine helps drivers see personnel and equipment in their vicinity.

[1]Similarly, the advent of more specialists in lubrication and oil recovery and recycling in the industry is likely to lead to overall reduced exposures to hazardous materials.

In addition to identifying technologies developed and used specifically to address occupational health and safety, discussants pointed to the health and safety implications of emerging technologies. Two representatives noted that the increased size of mining equipment—trucks in particular—raises health and safety concerns about vibration, visibility, and braking. In response, a large surface mine operation has outfitted each of its 240-ton haul trucks with three video cameras: one in front, one in back, and one to the right (opposite the operator's cab). Drivers use the cameras to scan for personnel and vehicles in their vicinity, as well as to position their trucks and monitor the dumping of material.[2]

Wheel-bearing-temperature indicators on haul trucks help operators avoid front-spindle failures. In a recent year, 40 trucks experienced such failures, with wheels coming loose on two occasions, reported one industry representative. Another posed the question of large-truck stability in case of a tire blowout. Devices that monitor tire pressure and temperature in real time are expected to reduce the incidence of blowouts, a source of occasional injuries and fatalities. Finally, two heavy-equipment manufacturers emphasized the importance of computer-assisted "fly-through" design tools that allow designers to virtually explore the operator environment, including visibility and access to controls, of potential equipment designs.

Mining-industry participants identified several promising technologies to improve health and safety that may be commercially available to the mining industry by 2010.

• *Vision assists.* Images generated using infrared sensors mounted on the front of a vehicle are projected in front of the operator. Such systems reportedly enable an operator to see four times farther at night and in poor weather.

• *Heads-up displays.* The technology to project control-panel information onto cabin windshields is being developed by the automotive industry and may be applied in mining vehicles as original equipment or as an aftermarket installation.

• *Proximity detectors.* Radar technologies for detecting nearby vehicles are reported to be in use at the facilities of two metals producers. The detectors "almost certainly" will be in common use by 2020, according to one discussant. Also under consideration are tags similar to theft-alert devices used in retail stores, which can detect personnel up to 50 feet away.

• *Voice-activated controls.* This technology currently is being developed by the automotive industry.

• *Biometric sensors.* In-cab imaging technologies are being developed that will monitor an operator's eye and facial characteristics for signs of drowsiness.

---

[2]While praising the potential value of cameras, several discussants questioned the reliability and usefulness of current technologies in the harsh mine environment.

Many important mining-equipment innovations raised in the discussions not only mitigate health and safety risks but also address the need to create a more enjoyable, interesting, and productive work environment. This is clearly important, given the need to attract and retain highly qualified workers and to maintain high productivity while operators are on the job.

Many pieces of mining equipment for both underground and surface operations are being designed with enhanced overhead protection and enclosed operator cabins. Several operating companies and equipment manufacturers cited the benefits of cabins for reducing exposure to noise, dust, heat, and vibration. One executive pointed out that over the past 20 years, construction of operator cabs has improved significantly, with better sealing and the increased use of sound- and vibration-mitigating materials. Increased cab leg room was cited as a notable amenity by one machinery-supplier executive. Joystick control schemes are also being increasingly incorporated into equipment. When applied to excavators and shovels, one executive observed, such features can offer important productivity benefits, since shovel operators often are seen as a key determinant of productivity at open-pit mines.

New generations of mining equipment feature enhanced noise and vibration attenuation, climate controls, ergonomics, and operator visibility.

• New-generation excavators situate the operator above the operating area, affording 180-degree visibility as opposed to the 90-degree visibility in older models. Dual cabs in some models add greater operator flexibility.

• An equipment manufacturer has introduced a doubly articulated LHD unit, with the cabin located on the central section. In addition to adding flexibility for working in tight environments, the design includes specialized suspension and ergonomic features that reduce physical stresses on the operator.

• One large mine operation reported working closely with an equipment manufacturer on enhanced noise-dampening involving mufflers, exhaust-system positioning, and blanketing.

TECHNOLOGY AND HUMAN-RESOURCES MANAGEMENT

A common argument raised by industry executives was that despite the prospect of automation and other technology enhancements, people are becoming *more* critical to the success of a mining operation, not less. Several relevant arguments were put forth:

• As mining equipment increases in scale and staffing levels decline, individual operators play a greater role in determining mine output.

- The mining workforce is aging, making retraining of older workers a priority.

- Achieving the productivity gains sought by both management and investors requires empowering the rank-and-file and upgrading their roles from following rules to solving problems.

As we saw in Chapter Four, IT is changing the way mining equipment is being deployed and operated. As mining equipment becomes more advanced through IT and communications innovations (sensors, advanced computing power, GPS, remote controls, and operator assists), line workers will have unprecedented access to information and control over the equipment they are operating. To maximize the productivity-enhancing potential of such technologies, miners will have to hone their multidisciplinary and critical-thinking skills. Moreover, operating companies will need to find or build expertise in areas such as computer programming, communications, and electronics. And these skills and capabilities will have to be refreshed and updated regularly.

"We must learn to adapt faster and faster, or we won't be able to stay in business. It will become increasingly important to make adjustments as close to the action as possible. Increasingly, supervisors and managers will focus their attention on providing information and knowledge so that employees can make the right decisions, rather than making decisions and directing employees' activities."

—*Mine manager*

Many operating companies and technology developers admitted that the interface between miners and IT needs to be improved. One technology supplier noted that it masks the complexity of its crusher control systems behind conventional, plain-looking control panels so as not to overwhelm operators. Another reported that the performance of his company's IT application at mines often deteriorated a while after it was initially installed, yet human factors were rarely identified by mining companies as the cause. The usual thinking is, "It's a computer, it should figure out the problem." When asked to identify the leading constraint on his firm's ability to boost productivity, one industry executive, effectively conveying the sentiments of several participants, declared flatly, "Getting people to think."

Not surprisingly, many executives stated that miners were their most critical asset. Yet competing for, recruiting, training, and retaining highly qualified, multidisciplinary workers in a strenuous "bricks-and-mortar" occupation presents a major organizational challenge—despite the high wages characteristic of the industry.[3] A coal-industry executive lamented that technological innovation in coal mining in particular was lagging because "the coal industry does not attract high-caliber, Ivy-League people." A manufacturer reported that the industry needed an additional 8,000 properly skilled people worldwide to maintain its machinery. Speaking on this point, a stone and aggregates executive reported that his organization faced a shortage of skilled workers and that this was having an impact on maintenance in particular. He worried that his equipment was "getting stretched out" as the firm attempted to keep up with growing demand.

> "We see mines doing all kinds of things. It really comes down to the individuals in leadership positions. Technology won't solve problems. It's a combination of technology and people."
>
> —*Technology supplier*

In approaching the subject of critical technologies, several industry participants explicitly downplayed the importance of hardware innovations in determining performance outcomes. A participant who is active across the coal industry asserted, "Productivity increases are being accomplished through operational changes—it's cheaper than the capital costs of technology." A representative of a major technology supplier identified people's attitudes and innovative mine processes as critical components of a successful mine, adding, "Technology is only one leg of a three-legged stool." Explaining how he doubled productivity and cut output costs by 40 percent, another participant from the metals sector observed, "We did it not by buying new equipment, but by motivating the workforce." He added that even when a new technology may be useful, building acceptance and a commitment to use the technology to its greatest extent ("getting people fired up") is an essential prerequisite to successful implementation.

In 1999, a major operating company adopted a business process-redesign program modeled after similar efforts at General Electric. The goal of the program is to "engage and energize" the workforce to improve productivity. The effort entails the formation of task forces to reexamine the effectiveness of all the company's operations and activities and to assess how they are integrated. A company manager described the goal of the program as "getting people to think, and then getting them to think together." The challenge of such corporatewide reengineering was not only that of improving productivity in a period of unprecedented competitive pressures, but also that of getting people to dedi-

---

[3]In 1997, mining employees earned the highest wages of all U.S. industry worker—nearly $44,000 per year, compared with an average of $29,000 for all industries.

cate themselves to mining in a new way in an age when the value of the industry in the United States has been open to question.

Several industry representatives noted that new-technology introductions require "buy-in" from the rank-and-file to be effective.

• At one facility, management introduced wireless radios to enhance communications in the mine, but workers resisted their use because of concerns about intrusive management oversight. Workers frequently failed to turn their radios on or reported that they were in "a noisy area."

• A representative from a research institution reported that prototype remote controls that he had installed on a company's mining equipment were summarily removed with an axe.

• Preventive maintenance requires that operators shut down and repair a piece of equipment well before a failure occurs. This can be seen as unnecessarily disruptive and threatening to immediate production targets. "Changing the way we've done it for the last 40 years," said one executive, "requires selling the program to the people."

Chapter Seven

CONCLUDING REMARKS

In the preceding chapters, we have presented executives' and managers' views on critical-technology trends in the mining and quarrying industry. In this concluding chapter, we build on these perspectives and draw several implications for public- and private-sector mining-technology research, development, and diffusion in the future.

THE DIVERSITY OF THE MINING INDUSTRY

Many of the technology trends presented in this report apply to all segments of the mining industry. Yet the industry is diverse, and there are some important differences between the major players and the ways in which they affect critical-technology trends.

The metals segment differs from other sectors in its strong dependence upon exploration and acquisition of the best ore bodies and in its greater globalization of operations. Metals producers thus spoke the most about international technology trends and lessons to be gained from abroad. Despite present low metals prices, product demand remains high, and hence increasing productivity and lowering cash costs remain fundamental goals. Metals producers tend to be investing more heavily than other mining-industry firms in complex and advanced technologies such as dispatch systems, high-precision equipment positioning, and super-size equipment. In the future, metals producers are likely to spearhead the implementation of advanced remote-controlled and automated equipment.

> "The primary concerns of the coal industry are regulatory reform, coal utilization, and energy policy."
>
> —*Coal-company executive*

Coal producers, on the other hand, presented themselves as less concerned about mine-productivity-enhancing technologies, primarily because of chronic oversupplies on coal markets and the resulting decline in coal prices (see Figures 2.2 and 2.3). As we note below, coal producers also are focused on major issues affecting coal utilization. Finally, historical precedent and tradition appear to be important influences in decisionmaking among coal companies, fostering a generally conservative outlook on technology prospects.

> "We have a clear strategy: to build on our existing strong market positions through investment and acquisition."
>
> —*Aggregates-company publication*

The stone and aggregates segment stands out for enjoying high product demand and prices, and hence it has a strong incentive to achieve rapid productivity increases. In addition, given aggregates producers' historical character as small-scale, relatively low-tech operations, product demands combined with a rapid wave of consolidation are resulting in significant investments to update excavating, crushing, and product-delivery technologies. Another important trend facing the segment is the increasing size and decreasing number of quarries, and their growing distance from urban centers. In sum, quarries in the United States are likely to see some of the greatest technology changes in the mining industry in the coming decades. As rising buyers of new machinery and equipment, stone and aggregates producers stand to become important drivers of mining-technology innovation.

The industrial minerals segment, while not widely represented in this study, generally enjoys the same favorable demand climate as aggregates producers. Interestingly, many industrial minerals producers appear to align themselves more closely with other industries, such as chemicals and agriculture. This may preclude important opportunities for collaborative R&D and problem-solving ventures with "mainstream" mining, as well as quarrying, firms and organizations.

RESPONSIBILITY FOR MINING R&D

Discussants from all corners of the mining community agreed that mining-technology R&D efforts in the United States have decreased substantially over the past few decades as a result of cutbacks in both public- and private-sector funding. As noted in Chapter Two, when we asked about their in-house efforts, nearly all of the operating companies stated that they had no formal R&D program. As a result, much of the recent innovation in mining has been in the form of incremental improvements to existing technologies. Very few participants mentioned any R&D efforts on breakthrough technologies.

Approximately 60 longwall systems were in use in U.S. coal mines in 2000. The demand for such mining equipment is limited, even as the technology becomes increasingly complex and R&D requirements rise. Photo courtesy of Joy Mining Machinery.

Contributions from universities, in particular, have dropped significantly, and two causes were cited during the discussions: First, university mining research is being perceived as too theoretical and not product-oriented enough to satisfy the needs of the mining community. One university participant mentioned his institution's efforts to counter this by forming a consortium that included industry and government partners, and by adopting corporate-style arrangements such as product milestones and deliverables. A second problem cited was the apparent mismatch between the duration of university research funding cycles and graduate-student tenures, on the one hand, and the increasing demand from commercial sponsors for short-term results, on the other. One university in our study had made a concerted and successful effort to adapt to this changing environment by taking on more smaller projects.

These circumstances appear to have resulted in a situation where, when a technology research interest is identified, it is unclear who within the community should champion R&D, demonstration, and commercialization. An example of this situation involves vehicle powerplants and the need for low-emission diesel engines and alternatives to diesel: Mining companies wait for truck manufac-

turers to address the need, and truck manufacturers wait for engine manufacturers; yet engine manufacturers may have enough business outside of the
mining industry to keep them from responding with the same sense of urgency
that the mining companies feel. Similar scenarios appear to exist for other
technologies, including alternatives to hard-rock blasting, remote controls, and
mineralogical sensors.

> "Mining is a very small industry with small numbers of [items of] equipment, but it has
> large research and development needs. What do you do? Risks and investment
> should be shared."
>
> —*Technology supplier*

Discussion participants repeatedly raised the concern that there was not
enough collaboration among technology suppliers and between technology
suppliers and operating companies. Collaboration, it was said, was needed
throughout the entire innovation process, from concept development to commercial demonstration. The need for collaboration is clear: The demand for
mining equipment and services is limited, advanced technologies are increasingly complex, R&D costs often are too great for a single company to take on
alone, and operating companies are wary of being early adopters. In addition,
greater collaboration between suppliers and operators will improve knowledge
flow and will facilitate both the relevance and the adoption of new technologies.
In short, R&D roles as well as funding levels need to be reappraised in the industry.

Research and development partnership funding provided by the DOE Office of
Industrial Technologies was mentioned as a positive step in this direction—
suggesting that the federal government can play an important role in convening
parties and catalyzing ventures. Several speakers also pointed to models of
successful, long-term government/industry collaborative research on advanced
technology development in Australia, Canada, and South Africa. As noted in
Chapter Two, several operating companies appear to be recommitting themselves to mining R&D, so the industry may be seeing a paradigm shift.

UPSTREAM AND DOWNSTREAM INNOVATION

When asked to highlight critical technologies, operating-company participants
often focused first on technologies in downstream activities (such as beneficiation and utilization) rather than those in upstream activities (such as ore extraction). Several study participants were able to discuss in detail the benefits
of monitoring and control technologies for optimization of their processing
plants, while the benefits of such technologies for optimizing ore production

were sometimes viewed as less critical or "too soon to tell." Similarly, when discussing activities at the mine site, participants tended to focus more on haulage than on development, drilling, or blasting.

This disparity partially reflects the fact that minerals processing operations—downstream activities—have more in common with factories and refineries, where process optimization technology has been in use longer, than do upstream activities. The bias toward downstream technology also can be understood from an economic standpoint for those commodities in which processing represents the bulk of the cost: The value of productivity gains tends to increase with the value of the product, and value is added as the product moves downstream through the various operations of a mine. One manufacturer estimated that a 1 to 2 percent productivity gain in a metals-processing plant was equivalent in economic value to a 20 to 30 percent productivity gain in an underground mining operation.

In the case of coal, producers and technology developers emphasized the importance of preparation plants and transportation: The industry is not production-constrained, and transportation can account for as much as 50 to 80 percent of the cost of coal. Thus, gains in product quality and transportation costs are usually viewed as more valuable than gains in extraction productivity. Downstream technologies also receive particular attention because of regulatory and community concerns. For example, technology to reduce emissions during coal utilization was commonly cited by coal producers as an important avenue to sustain the market for high-sulfur Appalachian coal and, more fundamentally, coal in general. This downstream focus of the mining industry can help explain the incremental pace of technology innovation upstream at the mine site.

> "The mining industry has ignored blasting, but improvements on the front end have a big impact on the back end."
>
> —*Technology supplier*

The incentive to innovate upstream operations may be increasing. As processing plants become more highly tuned—to meet higher productivity, emissions, or quality targets—the quality of feed materials becomes a more important determinant of plant performance. Similarly, the trend toward just-in-time delivery demands closer mine/plant integration to manage feed quantities. These two trends, in turn, are supported by the development of information technologies which are increasing the control over and the ability to link together unit-ops equipment. Drills and bulk explosives loaders, for instance, can be programmed to meet crusher demands. Finally, regulatory and community pressures (concerning aesthetics, noise, and land use, for example) increasingly challenge the basic character of mining operations, especially for aggregates, industrial minerals, and metals producers. This suggests that R&D and innovations targeted at upstream mining processes are likely to have higher payoffs in the future.

COORDINATION OF TECHNOLOGY AND REGULATORY DEVELOPMENT

An issue raised in the discussions was the desire among the mining companies for more coordination of technology development and regulatory development. The regulatory framework in some areas, such as ground control, was cited as being outdated and hence of limited relevance to emerging technologies. In other areas, most notably ambient noise and respirable dust, several participants cited compliance with existing regulations as being technologically impossible, and proposed new regulations raised additional concerns. These discontinuities have prompted calls for early, proactive technology-development efforts when regulatory revisions are considered.

> "Little forethought goes into co-developing technologies to successfully implement new regulations. The goal would be to have a technology available when a regulation is implemented, so as to provide for a smooth evolution for mining companies."
>
> —Coal-company executive

Further complicating the matter, claimed one participant, is the tendency for technology suppliers to focus on productivity-increasing and money-saving innovations, with technologies to address regulatory concerns apparently being viewed as unprofitable and thus unattractive for investment. More of the impetus for such development, he claimed, needs to come from the government.

> "The diesel technology is there for over-the-road applications, but engine manufacturers don't want to certify them for mines."
>
> —Coal-company executive

Finally, an opportunity for public-sector/private-sector cooperation to promote mining-technology development and diffusion is in certification. The mining industry can benefit from the application of many technologies, such as low-emissions diesel engines, that are presently or imminently available but whose technology developers do not see a strong business argument for going through the mining certification process.

WORKFORCE AGE

"The average age of a coal miner in the United States is 46 years. This miner is not getting any younger and, with age, is more susceptible to fatigue and injury. . . . It also takes longer to recover from any injury the miner does suffer. This reality provides an added incentive to reexamine both the work that we do and the way that we do that work."

—Coal-company document

In 1999, the average age of a coal miner in the United States was 47 years, and it is expected to increase in the future. The aging of the mining workforce was mentioned in several discussions, and the issue raises several important technology-related concerns: physical limitations of mine workers, ergonomics, know-how retention, and training and retraining.

An aging workforce in a strenuous industry raises questions about increases in injury rates, longer injury-recovery times, and more time lost due to non-mining-related medical concerns. Factors such as reduced stamina, flexibility, and strength need to be considered when designing equipment, planning maintenance, or setting protocols for mine safety and rescue procedures. Increases in chronic ailments such as back pain and deteriorating vision may also be important limitations. It therefore will become increasingly important to consider the needs of an aging workforce when developing mining technologies and operator interfaces—both to minimize injuries in the future *and* to enable mining and quarrying companies to retain their most skilled workers in a very competitive hiring environment. However, a coal-company executive noted that although he has tried to work with manufacturers, "not a lot of thought" has been put into designing new pieces of equipment with an older workforce in mind.

Mining skills are acquired largely through hands-on learning from experienced workers—especially in the area of safe practices. As experienced workers retire and as workforces are reduced in size due to productivity enhancements and consolidation, valuable knowledge may be lost. In addition, know-how retention becomes more important as mining technologies grow in complexity. Consolidation and globalization may alleviate this situation by shifting workers and sharing information among sites and companies. On the other hand, new technologies such as operator assists and remote controls that rely on the use of video images and joystick controls are readily adopted by younger workers, said several discussion participants. Research may be needed to sort out the implications for both productivity and safety of the aging workforce and technological change.

TECHNOLOGY CROSSOVER OPPORTUNITIES

Mining in the United States has a tradition of self-reliance: Mines typically are in remote locations, and the ore bodies they work often have unique characteristics. Yet, as discussed above, our discussions revealed a desire for increased information sharing and R&D collaboration in the industry. Executives from several of the larger mining companies noted that they hold periodic "summit meetings" among representatives from different sites to share information on operations and technology. As noted in Chapter Two, a few study participants had initiated efforts to identify potentially relevant technologies from outside industries. The National Mining Association also has a committee dedicated to technology issues. Nevertheless, existing information-transfer and technology-search mechanisms in the industry do not appear to be fully exploiting opportunities for technology crossover within companies, across the mining industry, or, importantly, with nonmining sectors.

Outside the mining sector, several technology areas may have valuable crossover applications. Previous RAND research has indicated that maintenance is a critical-technology concern for many industries. Diagnostics such as vibration, lubricants, and ultrasonic analysis are not yet used extensively in mining, even though these monitoring and analysis techniques are common in industries with much smaller capital investments in their fleets. Over-the-road trucking, construction, and manufacturing—like mining—have high downtime costs, and the mining companies may gain tremendous benefit from learning about new maintenance concepts and best practices in these sectors. Several other broad issue areas also offer technology crossover potential:

- As mines strive to reduce the cyclical or batch character of their operations, they can draw on technologies and know-how from refining and other processing industries. More generally, the mining industry can draw on concepts such as Total Quality Management (TQM) and "Six Sigma" enterprise process redesign (approaches developed and championed by General Electric and other manufacturers).

- Mining is not unique in having an aging workforce (although that workforce's extent and its implications may be more significant than in other industries), and the sector can learn from and share its experiences with other industries, such as those represented in the DOE Industries of the Future programs.

- Industries such as petroleum production and chemical manufacturing may have important lessons to share, for example, lessons on resource conservation and pollution prevention.

Of particular note are opportunities for technology and know-how crossover arising from the similarities between mining and the military in technology needs and applications and the working environment. Both sectors share a highly specialized and limited market for technology development; a reliance upon robust mechanical and electronic hardware that can withstand harsh operating conditions; and operational and logistical challenges imposed by remote, variable, and unpredictable working environments. Each sector offers complementary expertise that could benefit the other in areas such as position monitoring and wireless communications. The military can provide critical know-how to the mining industry in managing complex maintenance and repair operations; advanced use of modular designs, subsystem replacement, and centralized repair depots; and assuring rapid resupply of low-demand spare parts.

One potential solution is the creation of a communitywide "mining center of excellence" dedicated to generating and disseminating reliable and actionable information about technological needs and developments. The goals of a mining center of excellence might include

- Initiating innovative research projects aimed at attaining long-term strategic development goals in the mining industry.

- Acting as an information clearinghouse for technology availability, costs, case studies, and other relevant information.

- Stimulating strategic partnerships between researchers, suppliers, mining companies, and government agencies to bring together the complementary expertise necessary to address complex technological challenges.

- Providing individuals with the skills mix to become informed leaders in the mining industry.

MEASURING THE IMPACT OF NEW TECHNOLOGIES

One of the challenges of summarizing industry leaders' discussions is that of reconciling the widely differing views they expressed on the prospects for many new technologies, including the expected costs and benefits of expected innovations. These differences of opinion were rooted in varying assessments of the expected performance, costs, and benefits (the productivity value) of those new technologies. We believe that the variation in assessments is a result of a dearth of reliable information about costs, productivity benefits, and other consequences of technology investments, and the lack of information may be skewing priorities in favor of conventional technologies and practices and slowing new-technology development and diffusion.

> "It's surprising to find an operation that understands their costs from beginning to end."
>
> —*Technology supplier*

One explanation repeatedly cited was that operating companies—especially smaller ones—have limited knowledge of costs across their operation, often a result of the fact that mining and quarrying enterprises typically do not have sufficient financial and technical know-how to analyze such questions.

This makes technology assessment difficult. One manufacturer noted that even though the operating costs of his machinery were demonstrably lower (by more than an order of magnitude) than those of competing products, it was difficult for him to convince operating companies of his machinery's merit. Another executive, frustrated with the difficulty of convincing operating companies of the merits of his company's advanced analytic and problem-solving capabilities, inquired about lessons to be learned from other industries for better communicating technology benefits.

In addition to investment and operating costs, numerous other factors come into play in assessing a technological investment in mining. Among these are training and staffing requirements, mining productivity, energy consumption, environmental impacts, worker health and safety impacts, spin-up time, transportability to additional sites, and time to return on investment. Moreover, given the widely varying needs and capabilities of different mining operations, all of the above factors need to be assessed in terms of commodity type, mine size, mining method, geology, climate, local skill availability, community acceptance, and regulatory constraints.

Although suppliers provide new-product performance data to prospective buyers, several operating companies cast doubts on the reliability of such information, implying that it is especially inadequate for guiding decisions involving major investments and facility reengineering. Several reasons for this were offered in the discussions, including

- The limited number of technology demonstrations, such that there are often no installations in comparable operations.

- Insufficient resources to collect rigorous performance data.

- Insufficient communications, resulting in firms being unaware of applicable information regarding new-technology performance.

- A tendency in mining to assess effectiveness by experience and "feel" rather than quantification.

Finally, measuring productivity benefits of new technologies on parameters such as equipment availability and durability, reduced maintenance requirements, fit with existing systems and practices, and contributions to process integration and optimization are notoriously difficult tasks—for any industry.

Objective, third-party measurement and assessment of emerging technologies could better support critical-technology acquisition as well as R&D investment decisions. Such an effort would have the added benefit of providing a source of quality control for mining technology: Successful technologies could be highlighted, and unfruitful approaches could be identified and modified or eliminated. In addition, research in this area could help identify the key factors, e.g., company size or industry sector, that are important in making investment decisions. Finally, the federal government has a long-term interest in tracking productivity trends across the economy; better quantification of the productivity benefits of new technologies being adopted in the mining and quarrying sector would inform economic forecasting and monitoring efforts.

Over the past decade, the mining industry in the United States has shown greater productivity increases than other sectors such as manufacturing and construction. Yet many in the industry are concerned about mining's health and its long-term viability. Take the example of coal: Thanks to the introduction of new technologies, the number of labor-hours required to produce a unit of coal has dropped by a factor of eight since 1950, yet the industry suffers from overproduction, low market prices, and unfavorable profit margins. This suggests that conventional measures of success, which are geared toward perfecting methods of mass production, are no longer serving mining-industry decisionmakers well.

As we heard throughout the technology discussions, mining and quarrying operations today also compete on a range of other parameters: their ability to attract skilled and motivated workers; health and safety, environmental, and aesthetic impacts; supply-chain integration and getting products to the market at the right time; and, ultimately, customer and community satisfaction. These are critical measures that must increasingly be considered as new mining technologies are envisioned, developed, and assessed.

LIST OF STUDY PARTICIPANTS

COAL MINING

The American Coal Company
Robert E. Murray
Director

American Electric Power
David G. Zatezalo
General Manager, Windsor Coal Company

John L. Hamric
Manager, Mining and Project Engineering, AEP Pro-Serv, Inc.

Arch Coal, Inc.
Kenneth G. Woodring
Executive Vice President, Mining Operations

A.T. Massey Coal Company Inc.
Mike Bauersachs
Director of Acquisitions

CONSOL Inc.
Frank Burke
Vice President, Research and Development

Steven A. Cotton
Director, Mining Section

Interwest Mining Company
Ken Perry
Mine Development Administrator

Peabody Group
Christopher G. Farrand
Vice President, Corporate Affairs

David A. Beerbower
Vice President, Safety

Nickolas R. Kasperik
Senior Group Leader, Environmental and Quality Control

Jim Walter Resources, Inc.
Charles A. Dixon
Vice President, Engineering

Dale Byram
Manager of Safety and Training

METALS MINING

ASARCO Incorporated
Krishna Parameswaran
Manager, Regulatory Development

Barrick Gold Corporation
Gregory P. Fauquier
Senior Vice President, U.S. Operations

Canyon Resources Corporation
Rober D. Benbow
General Manager, CR Briggs Corporation

Cleveland-Cliffs, Inc.
Thomas J. O'Neil
President and Chief Operating Officer

Edward C. Dowling, Jr.
Senior Vice President, Operations

Echo Bay Mines, Ltd.
Donald C. Ewigleben
Vice President, Environmental and Public Affairs

Bill Goodhard
Director, Reclamation Environmental Affairs

Hecla Mining Company
Tim Arnold
Manager, Lucky Friday Unit

Douglas C. Bayer
Mine Engineer, Lucky Friday Unit

Homestake Mining Company
Walter T. Segsworth
President and Chief Operating Officer

Inco Limited
Greg R. Baiden
Manager, Mines Research, Ontario Division

Kennecott Utah Copper Corporation
Tom Probert
Vice President/General Manager, Business Improvement

William R. Williams
Vice President/General Manager, Technical Services

Newmont Mining Corporation
Larry Clark
Consulting Engineer, Technical and Scientific Systems

Kim Eccles
Chief Mine Engineer, International Operations

Philip Walker
Manager, Metallurgical Services

Lance N. Throneberry
Manager, Worldwide Agreements

Phelps Dodge Mining Company
John O. Marsden
Vice President, Technology

AGGREGATES AND INDUSTRIAL MINERALS MINING

Rio Tinto Borax
Preston S. Chiaro
Chief Executive Officer

Paul J. Zerella
Chief Technology Officer, U.S. Borax, Inc.

Rodney Drake
Mine General Supervisor, U.S. Borax, Inc.

Hanson Aggregates East
David Lange
Vice President, Engineering and Technical Services

Martin Marietta Materials, Inc.
Stephen P. Zelnak, Jr.
Chairman, President, and Chief Executive Officer

Vulcan Materials Company
Guy M. Badgett III
Senior Vice President, Construction Materials

EQUIPMENT AND SERVICE PROVIDERS

Atlas Copco Wagner Inc.
Casper Swart
Manager, New Product Development

Charles R. Chelin
Business Line Manager

Austin Powder Company
David P. True
Vice President and General Manager

Bucyrus International, Inc.
Greg G. Dinkleman
Vice President, Engineering

Caterpillar Inc.
Denis J. Mills
Business Process Manager, Corporate Mining Group

Richard R. Burritt
Manager, Mining and Earthmoving Technology Systems, Corporate Mining
Group

Dyno Nobel Inc.
E. J. Burke III
Leader, Distributor Sales Industry Management Team

Lawrence J. Mirabelli
Leader, Technical Services Core Team

ExxonMobil Lubricants and Petroleum Specialties Co.
Tom Hennessey
National Account Executive, Commercial Vehicle Lubricants

Fosroc, Inc.
Alan A. Campoli
Business Development Manager

Goodyear Tire and Rubber Comapny
David L. Wright
Chief Engineer, Off-Road and Farm Tire Field Engineering

HLS Hard-Line Solutions Inc.
Walter Siggelkow
General Manager

Jennmar Corporation
John Stankus
Vice President, Engineering, and President, Keystone
Mining Services

John T. Boyd Company
James W. Boyd
President

George V. Weisdack
Vice President

Thaddeus J. Sobek
Senior Mining Engineer

Charles H. Wolf
Senior Environmental Specialist

Joy Mining Machinery
Larry R. Buschling
Vice President, Global Engineering

Lynn Wheatcraft
Director, Application Engineering, North America

Kennametal Inc.
James Courtney
Manager, Sales and Marketing, Global Mining and Construction

Komatsu America Corp.
Edson R. McCord
General Manager, North American Research and Development Center

Liebherr
Francis Bartley
Manager, Research and Development Group

MAPTEK/KRJA Systems Inc.
Barry Henderson
Vice President, Sales, Marketing, and Administration

Master Builders, Inc.
Michael Rispin
Mining Manager, Underground Systems

Matheson Mining Consultants, Inc.
Colin Matheson
Mining Engineer

Mining Technologies International
Vern Evans
Group Technical Services Manager

David Ballantyne
Program Manager, Mechanized Drilling and Automation

T. C. Storey
Product Manager, Rail Haulage and Utility Vehicles

Donald Clark
Product Manager

Modular Mining Systems
James Wm. White
Chief Executive Officer

Mark Baker
Executive Vice President

NSA Engineering, Inc.
Joel A. Strid
Executive Vice President

Orica USA Mining Services
Fortunato Villamagna
Director of Technology

P&H Mining Equipment
Barry J. Turley
Vice President, Project Management

Robert H. Quenon
Mining Consultant
Chairman Emeritus, Peabody Holding Company, Inc.
Director, Newmont Mining Company

RAHCO International
Richard W. Hanson
President

Sandvik Tamrock LLC
Jean-Guy Coulombe
President

Paul L. Painter
General Manager, Mining and Construction

Trimble Navigation
Mark Nichols
Senior Director, Mining, Construction and Agriculture

Varis Mine Technology Ltd.
Matt Ward
Brian Falter

RESEARCH AND GOVERNMENT ORGANIZATIONS

Carnegie-Mellon University
Robotics Institute
Timothy E. Hegadorn
Research Engineer

Colorado School of Mines
Department of Mining Engineering
Tibor G. Rozgonyi
Professor and Department Head

Ugur Ozbay
Professor

MIRARCO/Laurentian University
Peter K. Kaiser
President

Pennsylvania Department of Environmental Protection
Robert C. Dolence
Deputy Secretary for Mineral Resources Management

University of Arizona
Department of Mining and Geological Engineering
Paul J. A. Lever
Department Head and Associate Professor

University of Utah
Department of Mining Engineering
Michael K. McCarter
Professor and Chairman

Virginia Polytechnic Institute
Department of Mining and Minerals Engineering
Roe-Hoan Yoon
Director, Center for Coal and Minerals Processing

Tony Walters
Associate Director, Center for Coal and Minerals Processing

Gregory T. Adel
Professor

Gerald H. Luttrel
Professor

Erik Westman
Assistant Professor

Christopher Haycocks
Professor

DISCUSSION PROTOCOL

BACKGROUND

1. Please describe your firm's mining-related activities. Is your firm executing or planning a shift in strategy?

2. What are the most important market and management trends governing your market segment? How are they affecting your firm?

3. In relation to its competitors, how does your firm excel or stand out in terms of the mix of technologies deployed?

CRITICAL TECHNOLOGY TRENDS AND DEVELOPMENTS

4. Considering the issue of mine productivity, what are the most critical factors (bottlenecks) you are tackling to improve productivity?

5. What are the recently available (or currently emerging) mining innovations critical to the success of your firm (or to the mining companies with which your firm works)?

6. What impact do these mining innovations have in terms of

    Enhancing productivity?
    Developing new products/markets?
    Opening up new reserves/extending existing ore bodies?
    Health, safety, and the environment?

7. Are the innovations part of a broad trend in the mining sector?

    What kinds of firms (facilities) are adopting these technologies (e.g., big/small, new/old)?
    What is their state and rate of diffusion across the sector?

8. What longer-term technological and management breakthroughs are likely to make the greatest difference to your firm and, more generally, to the mining industry?

THE INNOVATION ENVIRONMENT

9. What factors impact technology innovation and diffusion in the mining industry (e.g., R&D, commodity prices, consolidation, globalization)?

10. What is your assessment of the character and source of innovation in your segment?

    Incremental gains or breakthrough advances?
    Spinoffs from other industries?

11. Looking at the stream of technology/practice innovations you have identified as critical, where are they coming from (e.g., technology users/suppliers, in-house R&D, research institutions, abroad)?

12. How do health, safety, and environmental regulatory drivers governing the industry affect critical-technology changes?

13. The market for mining technologies is becoming increasingly global. How does globalization affect technology trends in the U.S. mining industry?